S.T.O.P.

The Project Management Survival Plan

A proven project management framework complemented with tools, tips, and techniques to ensure your success in project delivery and guarantee you're not the reason for *stopping* the project.

STEVEN STARKE

What People Are Saying About This Book

"The true value of project management is so much more than just organizing and managing a bunch of tasks and activities. In *S.T.O.P.*, Steve reminds us of the real value-added role a project manager plays within an organization and provides a strategic and pragmatic framework to successfully lead a project team through inevitable obstacles to completion — a must read for any project management professional."

— Bryan Williamson, Vice President and General Manager, ecommerce at Claire's Stores

"*S.T.O.P. — The Project Management Survival Plan* is an essential part of any project manager's tool box if they want to truly make a difference within their company and in their career. The book is chock-full of original — as well as best-practice — tools, tips, and techniques, from a variety of methodologies, that are artfully put together in real-world examples to demonstrate how project managers should think critically about delivering value."

— Tom Corcoran, Director of Program Management, Thomson Reuters

"Finally, a departure from 'schedule, scope, and cost,' as the mantra for technology-focused project managers. Steve's focus on project success criteria and business outcomes is what is needed in today's fast-paced technology-driven era."

— George Moore, Chief Technology Officer, Elsevier Health Sciences

"At the core, the S.T.O.P. principles should serve as the foundation for all professionals in today's fast-paced business environment. Many different areas of business would benefit from them, not just project managers. Steve Starke does an outstanding job of blending real-life examples of his experience and others with the fundamentals of S.T.O.P. and how to apply them. A great read for all levels of the organization."

— Dan Dierking, PMP, Chief Information Officer, National Council of Real Estate Investment Fiduciaries (NCREIF)

"Steve provides a candid look into his experiences and lessons learned in a way that will help every project manager approach things differently the very next day."

— Matt Jackson, PMP, Founder, Jackson Consulting

"*S.T.O.P.* is an invaluable survival guide. The principles outlined in *S.T.O.P.* provide the project manager with the necessary skills to organize a team, effectively plan, and manage the most difficult projects. Steve takes the complex discipline of project management and provides a common sense approach for resolving those challenging problems that arise during the course of a project. I am positive that the tools and techniques throughout this book will help you successfully manage the right projects and more importantly, *Stop* those projects that do not bring value to your organization."

— Brian C. Case, PMP, Director of Program Management, Fenwal Inc.

Author Online at
www.ActuationConsulting.com

ISBN: 978-0-9831116-8-9
Library of Congress Control Number: 2011917535

Project management
Program management
Project management — Handbooks, manuals, etc.
New product development

www.ActuationPress.com

Acknowledgements

— To my loving wife Lan for her inspiration, understanding, and support. The fact that she loves me must mean there's hope for me yet.

—To my children Mia and Tyler — dream big, create even bigger!

— Thanks to Greg Geracie for his guidance, direction, and friendship.

— A special thanks to all the peer reviewers for their time and effort in helping refine this book. Specifically, Brian Case, Bill Needham, George Moore, Alvin Doss, Tom Corcoran, Matt Jackson, Bryan Williamson, Dan Dierking, and Kjersten Moody.

— Kathleen Riley for visual creativity and direction. She gave this book life with her images and drawings.

— Beth Waibel for all her copyediting and proofreading work. She allowed me to look much smarter than I really am.

— Nina Villacci from Bellafly Design for book layout and structure. Thanks for dealing with my perfectionist side.

— James Steidl from James Group Studios, Inc., for taking an idea and turning it into the coolest book cover ever. Thanks for standing in front of that train. Sorry about the ticket, Jamie!

— Doug Wilkerson and Jeff Barningham for providing me the opportunity to earn my chops back in the day. Thanks for taking the chance on me.

— Robert and Gloria Starke, for their spirit, guidance, support, direction, and love. Thanks Mom and Dad!

— Mary Caronna, your memory lives on.

TABLE OF CONTENTS

INTRODUCTION

Stop for a moment and ask yourself, "Why do so many projects fail?" When is the last time, you, the project manager, yelled *Stop*!? As a project manager, have you lost control of your projects? Are your projects run-away trains headed off the tracks before they even start? The environment that project managers now operate in has turned global, with faster "to-market" turnarounds, higher sensitivities to regulatory compliance, and zero tolerance for low quality. Platform solutions, rapid advancements in technology, and offshore resourcing make projects and their path to success more complex, tortuous, and fraught with risk.

Today's project managers often have more questions than answers. Their attention has turned inward – with a heavy focus on process and the proliferation of different development methodologies. This in turn takes their focus off the big picture, success criteria, and overall value their project brings to the organization and customers. I believe this shift in focus is causing more projects and products to fail.

Even if the development methodology is executed flawlessly, it doesn't mean that the project – and ultimately the product – will be successful. Project managers seem to be standing at the crossroads – trying to determine when to go deep into the tactical elements of the project and when to keep a broader, more strategic focus on the bigger picture. This conflict, coupled with increasing organizational complexities and weak matrixed functional alignments, has created conflicted environments for project execution. Has this conflict caused you to lose the executive support and understanding you need to be successful? Is this the equation that stands before you, defining your projects?

Technical complexity + Lack of strategic direction + Weak matrix + Siloed environments = Loss of executive understanding and support

To effectively lead a cross-functional team in the delivery of a product, project managers must have a crystal clear understanding of the project's purpose, scope, and success criteria. This is especially true when dealing with a family of products, where the team is supporting multiple project streams with various interdependencies. The team needs to rely on the project manager to understand and clearly articulate the boundaries of scope, the constraints on budget, delivery timelines, and delivery process throughout the product's development.

To adapt to this increasingly complex environment, project managers must revise their core principles of operation to meet the new demands of their teams and organizations. Project managers must now have more robust recommendations and better answers for planning, execution, delivery, and overall management responsibilities of the assignment. They must now come to the table with not just a project plan, but a project survival plan, enabling skillful navigation around anticipated and unanticipated obstacles.

How do you learn to build a survival plan? There are plenty of management books; but the project management field especially is so saturated that it's easy to experience information overload. Sorting through the volumes of available information to find practical nuggets of survival wisdom as you navigate through your project is time consuming, and time isn't a luxury project managers typically have.

I learned two very valuable lessons early in my education that helped me survive project management and succeed in today's organizations. Oddly enough, they didn't come from my on-the-job training, but from residual lessons that stand out now that I understand how to succeed in corporate environments. These lessons came from unexpected places. The first was a scuba diving class. It took me about five classes to figure out what the instructor was really teaching. He wasn't just teaching us to scuba dive; he was teaching us how to survive while we were scuba diving. This concept became very clear when he made us take off all of our scuba gear, tie it up, and throw it in the bottom of the pool. The test was to jump in, untie the gear, put it on, and start scuba diving — all without coming up for air. Fun, right? The scuba instructor always said, "You never know what kind of situation you're going to be in. My job is to teach you how to think and handle yourself in any situation."

The second unconventional lesson I remember was from a college calculus class. It wasn't really the material but something the instructor said that made an impression. He started class one day by saying, "I'm going to give you the one secret to life today." Obviously, that got our attention. He

> If you want success as a project manager, you must combine project management methodology with an ability and plan to survive. Both are important, and neither can exist on its own!

said, “The one secret to life is... think clearly. If you can do that, you can solve almost any problem.” When faced with a threatening situation, if you can think clearly, you’ll find yourself managing relatively easily and successfully. But is it that simple? That to be successful in your projects all you need to do is think clearly and persevere? Is this the equation?

Survival plan + Critical thinking = Best chance for success

Survival Skills

One of the basic definitions of survival is *the development of techniques and tools a person uses to survive a potentially dangerous situation for an undetermined time. These techniques are meant to provide the basic necessities for life: water, food, shelter, AND the need to think straight, allowing you to navigate safely and avoid unpleasant interactions.* As I worked through how the marriage of clear thinking and survival skills could possibly have anything to do with project management survival, I was struck by yet another early life lesson, this one from the Boy Scouts of America. Now, before you close the book or send me a referral for a psychologist, stick with me for just a few more lines and you’ll have the fog lift — just like I did when I was searching for a practical way forward.

The Boy Scouts use a mnemonic device, S.T.O.P.[1], to address the mental aspects of survival. S.T.O.P. stands for *Stop, Think, Observe,* and *Plan.* Ironically, it was based around the same four-letter word I found myself wanting to say as I witnessed more and more errant projects. You see, being a project manager in today’s environment really means operating in survival mode at all times. You never know when some

[1] Boy Scouts of America. *BSA Fieldbook.* Irving, TX: Boy Scouts of America, 1984.

unforeseen circumstance or unexpected surprise from a team member will suddenly send your project off the rails. The more you're mentally prepared to survive, the more you're likely to do so. In project management, it seems that you're always dealing with emergency situations. You must have the ability to be decisive, think on your feet, and remain calm at all times. When your project deviates from the original plan, you need to be proactive and prepared, not simply reactive. Losing your cool isn't an option.

A project manager must also know when to recommend *stopping* the project. Stopping isn't something a lot of organizations do well. In a recent study, the Accept Corporation and the Association for International Product Marketing and Management (AIPMM) found that more than 60 percent of executives say they struggle making kill/go decisions.[2] There is a tendency to continue projects and activities even when most people involved realize it's not an optimal use of their time. Organizations generate a cultural momentum that, like a battleship, won't turn easily or quickly even when everyone is aware of the issues. The Boy Scout's S.T.O.P technique definitely seemed applicable to these types of circumstances. It isn't just a four-letter word on a street sign or something that's useful for survival, but a fundamental management principle to use in becoming a successful leader. It is a strategy to be used for project management and corporate survival.

[2] Accept Corporation and Association of International Product Marketing and Management (AIPMM). PPM is Dead. Long Live Portfolio: Q2-2011 Study – Portfolio Management Priorities. June, 2011. Retrieved from accept360.com.

Here are the fundamental principles of S.T.O.P.:

- **S**top before you act. Don't rush into a situation without considering the risks and the benefits.
- **T**hink about what you're about to do. What's the right way to safely accomplish the task?
- **O**bserve the situation and surrounding environment. What are the risks? How can I reduce them?
- **P**lan to reduce the risks and decide how to best implement the plan with safety and transparency.

Stopping and thinking before you act is just good common sense. How can you argue with that? But what makes S.T.O.P. really powerful is the "O," *observe*. Being able to observe your surroundings and the people around you gives you further context and a better understanding of your situation. This means not just reacting to a *current* situation, but observing situations *over time* and collecting enough data points so you can assess a new situation with statistical significance. Recognizing the need for data before making a decision is extremely important because it forces you to ask questions before acting. Data and observation give you the ability to think logically and clearly, forcing a thoughtful process that hedges against the far-too-common urge to look busy. Data and observation allow you to respond intelligently to problems and prevent you from responding with, "Data?... Um, no thanks. I don't have time to think things through. I'm a busy project

> An analytical approach, considering both the qualitative and quantitative aspects of a project, is vital to developing a survival plan and best positions a team for success.

manager and people expect me to know the answer, take charge, and keep moving."

What I also like about the S.T.O.P. principle is that it's tied to a deliverable. The "P" stands for *plan.* A plan, an actual tangible asset, thought through with a clear head and ready for reference when the waters get choppy or a fire breaks out. A plan designed to initiate and direct action before that action takes place. A plan allowing you to truly *manage* the project. When I discovered this part, I knew I was onto something, because this was missing from a lot of the other concepts and methodologies I had found. Stopping and putting a plan together sounds so basic, but I've found that it's often forgotten or overlooked. And what sounds so simple could be the furthest thing from your mind when you're battling a blaze. That's when the principles of S.T.O.P. can be applied.

What Does It Mean to "Manage"?

The word "management" sometimes confuses people. For some reason they focus on the word as a noun, as in a group of people, when in fact, they should also consider it as a verb. *Management* is a series of actions to monitor and control an environment or initiative. *Managing* is about context and understanding what the ideal end-state will be and putting a plan together that allows you to get to that end-state, while dealing with change that may occur along the way.

This all-purpose understanding of what it means to "manage" is only generalized in most project management literature. In fact, even project management standards, such as *A Guide to the Project Management Body of Knowledge (PMBOK©)*[3], only briefly touch on the concept when they recognize that

[3] Project Management Institute. *A Guide to the Project Management Body of Knowledge: PMBOK Guide, 3rd edition.* Newtown Square, PA: 2004.

general management skills are typical of those needed by a project manager. But they don't go into specifics regarding its application. If you think of general management as encompassing strategizing, planning, organizing, staffing, executing, delivering, and providing the proper infrastructure associated with support and operational processes, and you think of project management as the act of managing change against clearly defined project success criteria, you can see how project management is essentially general management for a project or program. But can't we simplify this concept even more by saying project management is essentially the ability to manage *value, team performance,* and *communication* in relation to the project?

Let's revisit the S.T.O.P. principles and tailor them so they're relevant from a pure project management perspective. These principles will be the framework we build on throughout the rest of the book.

- **Stop** before you act. Don't rush into a project without considering the risks and the benefits. Treat your project like a mission, where success is clearly defined and the only option. You're a valuable member of the team and they're counting on you to manage them through adversity and lead them to success.
- **Think** about what you're about to do. What's the right way to best accomplish the project? Make sure it's aligned with your sponsor's goals and expectations. If those goals haven't been defined, make sure they get defined before leading your team into battle — I mean, into the project.

- **Observe** the situation and surrounding environment. What are the risks? How can you reduce them? Put together a risk management plan that's relevant to you, your team, and your sponsor.
- **Plan** to reduce the risks and decide how to best implement that plan with quality and transparency. Put proper feedback mechanisms in place so you can monitor the project continuously and constantly look for ways to improve.

S.T.O.P. is easy to remember and can be applied to just about any situation. The goal of this book is to demonstrate the value of this principle by applying it to several different core areas of project management. You'll learn how to apply S.T.O.P. and, in the process, gain the foundational skills and understanding to become a better overall project manager and contributor to your team and your organization.

The next section will focus on using the S.T.O.P. principles to address some basic components essential for running *any* project. This book is designed to provide the essentials you need to be successful as a project manager — even if you have never managed a project. For experienced project managers, these next sections are meant to challenge your current thought process and introduce enhanced critical thinking skills as you interact with your teams, ultimately providing better results and more confidence for achieving success.

This Book's Mission

If you think about your project as a survival mission in which it's your responsibility to save and lead members of your team to safety, you may change your approach to managing your projects. You may start to understand your team members better: their strengths, weaknesses, and the environment in which they perform best. You may also have a new sense of urgency considering what's on the line. Many times on survival missions you only have one shot to get it right. You could say the same thing about planning and managing a project. If you don't get it right up front, you may find yourself and the team in front of the CEO explaining the failure, missing customer expectations, and hurting your career. With the role of the project manager, there's a lot at stake and no room to take chances. Corporations are counting on project managers more than ever to help them navigate these tough decisions. You need to rely on your tools, team, and skills. You need S.T.O.P. (Stop, Think, Observe, and Plan).

The S.T.O.P. management process is designed around the survival principles taught at the Boy Scout level and enforced by many of our armed forces divisions. The S.T.O.P. principles were designed to ensure that critical thinking occurs in high pressure situations (sound familiar?). I have taken those same principles and constructed a framework for project management planning, execution, monitoring, and control.

The framework is designed to ensure you've done everything you need, from a project management perspective, to ensure that you and your team succeed. There are no short cuts, workarounds, or silver bullets. If you think you want to cut

> The act of stopping projects doesn't come easily to most organizations. It requires a paradigm shift in understanding that stopping the project is a good thing. The S.T.O.P. principles will help facilitate that cultural change.

corners, be prepared for those cuts to haunt your project later. If you think what worked on your last project is going to work on your current one, be prepared for rework, missed dates, and dashed expectations.

Every project is unique; that uniqueness requires adaptability and the willingness to acquire context. What you need is an adaptable, focused, efficient, and smart way to manage your team and ensure that value is delivered to your organization. This book is all about helping you find that "way."

Each section of this book builds upon the next, walking systematically through the S.T.O.P. principles for each aspect needed to manage a project. Examples are the best way to learn, so I provide as many real-life stories as possible to give you further context and understanding of the material. At the end of each section, I summarize the important information, making it easy to find and reference later. Where appropriate, I use tables and pictures so that you can add them to your toolbox and reference. By the time you finish reading this book, you should have the tools you need to create a clear path and plan for how to proceed and succeed on your next project.

Over my career, these principles have proven to be invaluable in creating, directing, and managing teams and projects to success. My goal was to provide you, the project manager, with the elements, context, and tools you need to create a consistent, repeatable, and adaptable approach to designing a project plan that yields results. Not all projects are destined to be completed; *some need to be stopped.* A project stopped for the right reasons should be considered a success. I believe I've laid out a process that will ensure that — if your project is viable — it completes successfully. AND if it's *not* viable, you have the data you need to stop it and stop it quickly!

SECTION ONE

Value

S.T.O.P. and Figure Out What "Success" Means

One of the first skills required in project management is the ability to stop before acting. This comes by understanding the goals and success criteria of the project. Taking the time to construct well-thought-out project success criteria with your team will force them to evaluate their current actions and ensure that the plan moving forward is sound and in line with stakeholder expectations. One way to collaboratively construct success criteria with your cross-functional team is through project definition sessions. Project definition sessions are interactive discussions where the team and the stakeholder define the goals of the project and flesh out any assumptions, constraints, or expectations that need to be incorporated into the project plan.

Facilitating project definition sessions and helping your team understand their project success criteria is one of your main responsibilities as project manager and the first true value-add activity you can give to your team.

Project success criteria must be understandable, measurable, and achievable by the project team, stakeholders, and sponsor. Below, we tailor the S.T.O.P. principles and apply them to defining project success criteria:

- **Stop** before you act. Don't rush into a project without fully understanding the business drivers, goals, and how each will be measured and translated into success criteria.
- **Think** about what you're about to do and make sure it's aligned with your sponsor's goals and expectations. If

those goals haven't been defined, make sure you define them immediately.

- **Observe** the situation and surrounding environment. Do all of your stakeholders and sponsors agree with the success criteria you've defined? Has the project been given sufficient resources to succeed? Is the timeline realistic? Is the organization committed to the project?
- **Plan**. Develop your plan to define and measure the success criteria. Put the proper feedback mechanisms in place so you can monitor the project's success criteria continuously, looking for ways to mitigate risks where appropriate.

The project's sponsor may not know or need to know all the details of the project's scope. The early stages of a project, especially the conceive/market definition phase, can be a bit fuzzy regarding scope. However, the sponsor should know what's going to define value and success. As the project manager, you need to understand what success means to your sponsor. You must capture the requirements that define success so you know it when you see it. What metrics are the team going to be measured against once this project is complete? What truly defines the overall value of the project? Make sure you write that criteria down — and then show them to your sponsor. Ask, "Is this what you said?" If it's not measurable or clear, keep asking analytical questions until you can either make it measurable OR you can discard it.

If you don't gain understanding on this part of the project up front, the rest will be riddled with surprises and unknowns. You'll never be able to manage your sponsor's expectations or achieve success. Success that's well defined, agreed upon, and frozen will provide you that opportunity and a clear path to actually achieving it.

Remember that each criterion, ideally, should have a customer who will benefit from it. That could be an external customer for whom workflow or usability is enhanced, or an internal customer for whom performance is improved. Understanding the customers of the project is critical to comprehending what will make the project successful. You must understand *who* is going to benefit from *what* and *how* they're going to measure those benefits. Find out what their pain points are and what results they expect.

> So often, after being assigned a project, project managers try to run before they walk. This is especially common when the project is already in progress. You can quickly get caught up in the momentum of work and forget to question whether the work is justified. Talking with your sponsor is one of the first ways to stop. Stop yourself and allow both you and your sponsor time to think about the project. Many times just asking basic questions and allowing your sponsor to hear their answers out loud will give them the ability to assess the project and ensure it's still viable or determine whether it should be stopped.

Success criteria are typically focused around the areas of scope, schedule, and cost. They are detailed statements structured around *what's* going to drive direction and value of the project. They should not describe *how* the success criteria will be delivered. Examples may include that the project is delivered within ten percent of its original budget or that sales tools and demo data are considered a part of the scope of the project and are created within the product to ensure the sales team has what it needs to sell the solution.

The scope of the project is really what's going to drive the business value of the initiative. Figure 1.1 is a table you can use to capture your project success criteria. Something very simple that captures the "who, what, why, when, and how" of each success criterion and how it will be measured is all you need.

Figure 1.1 Success Criteria Table

	Title (What?)	Objective (Why?)	Description (How Will It Be Measured?)	Stakeholder (Who?)	Deadline (When?)
1					
2					
3					

Once you've captured your success criteria, make sure you post it in the workspaces of your entire team and your sponsor. This is what will define your project's mission and give your whole team the ability to manage your customer's expectations appropriately. I've found the checklist in Figure 1.2 extremely beneficial in determining whether you've done a good job capturing your success criteria.

> Remember, good project managers know that project success is not whether it's delivered on time and within budget, but whether it delivers value and meets the defined success criteria.

Figure 1.2 Success Criteria Evaluation Checklist

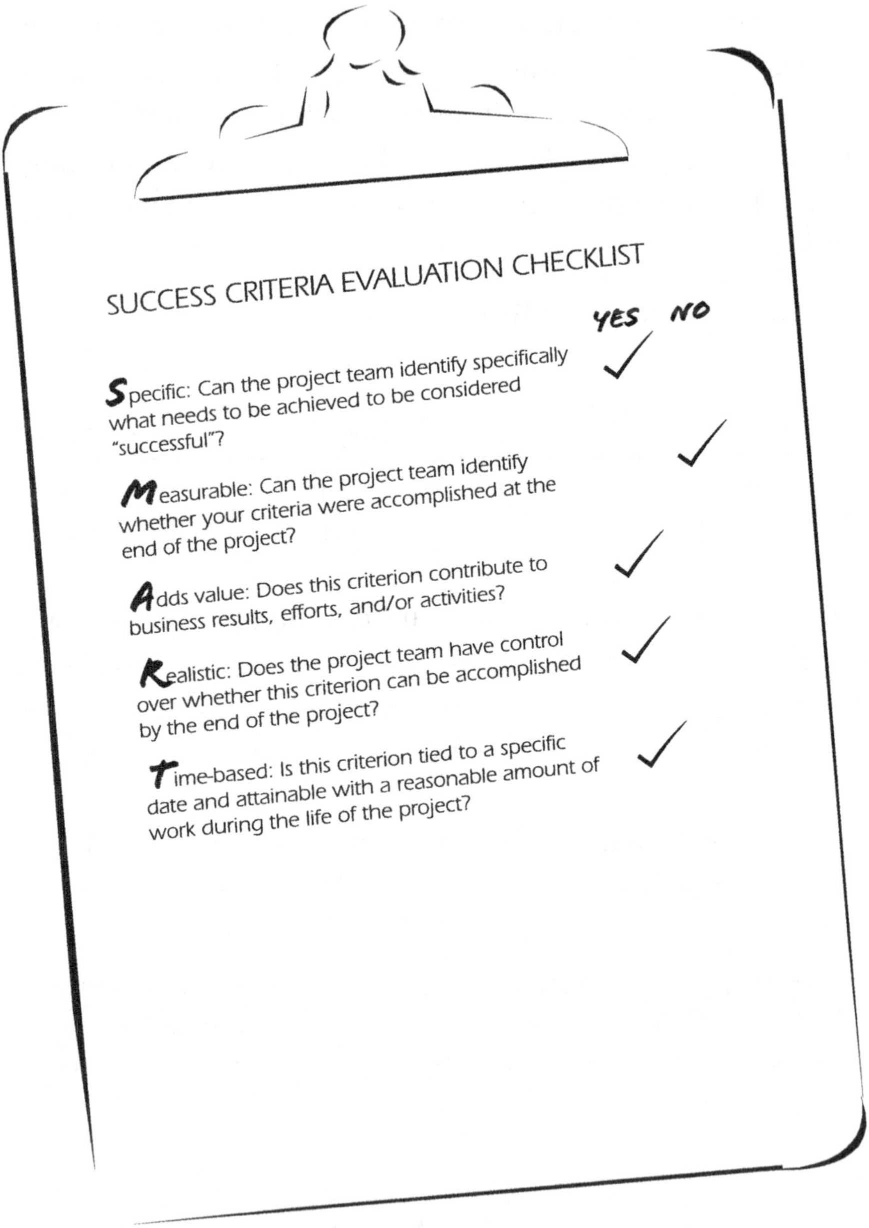

> A project schedule that's rolled up by success criteria and business objectives acts as that communication tool between business and technology, positioning you as a liaison between business and development and keeping everyone on track. This type of project schedule also serves as a proverbial stop sign for your sponsor. Once your sponsor can relate the work to the value, they can better assess whether that perceived value is attainable or if the value has been degraded by elements of time and cost. If relative cost is greater than relative value, that portion of your project should be stopped.

Now, how do you take capturing success criteria to the next level? What's the next step? So often I hear that the product manager or the business leader looks at the schedule, laughs at the project manager, and says, "What am I suppose to do with this?" indicating that the schedule has no value or relevance to them because it's too hard to read. You can make your project schedule more meaningful to your business sponsor by adding a column that allows you to map business success criteria to project-level milestones.

Essentially, what you're doing is giving your project schedule a fresh perspective — one that your sponsor can relate to. By incorporating success criteria, you allow the schedule to be rolled up and viewed by success criteria. You can then track and communicate progress in language that your sponsor understands. Establishing this link is critical in taking that next step.

The business and development teams may not understand all the technical detail or lower-level task information, but they will understand the mapping of the milestones to success criteria. This technique will help them understand why they should care and comprehend the associated task dependencies. You'll be amazed at their increased interest in your schedule and how they'll actually start using it and not just relying on you to report the information to them. It's all about getting and keeping your sponsors engaged by providing true value and relevance in a language they can understand.

The Value Equation of Stopping a Project

Once you establish your success criteria, you essentially have to relate or categorize each item to scope, schedule, or cost. Some may argue that an additional constraint of quality/ performance should be added. However, my opinion is that quality and performance should be considered *within* your scope and not as a separate category. Once you categorize your success criteria, you're ready to balance and manage your project. Basically, every project will be a balance of this infamous iron triangle — scope, schedule, and cost. The project tradeoff matrix will help you balance the three. However, to really utilize the triangle, you must understand what you're balancing against and what you're truly asking your sponsor to trade off. Let me explain.

You're raised as a project manager with a visual implanted in your head of a triangle that flexes and constrains as you move through your project. You force your sponsors to make decisions and tradeoffs against the three, but based on what? The answer should be value. When creating, developing, and delivering a product to the market, you seek to maximize its value. You want to maximize the customer benefit and experience while maximizing the return on investment. When you adjust your perspective around the triangle like this, you realize it's more of an equation based on deriving value. I call this the *project management value equation*. It's designed to give context to "scope, schedule, and cost," ensuring that you're weighing all that you do against the overall value of the project and keeping your sponsor and team focused on the prize.

Said another way, it's an equation meant to quantify and assess the value of a project and identify — if value has been decreased — whether the project should be stopped. The equation is:

$$Value = Scope \div (Schedule + Cost)$$

By understanding this concept, you bring more depth and meaning to what you really need your sponsors to trade off against. By assigning a value to each of your success criterion, you in essence are quantifying the value of the project. Increases to schedule and cost will decrease the value of the project if more scope is not being delivered to offset it. Ensure that your scale for scope, schedule, and cost are the same. Assigning either a percentage value or using a 1–10 scale can work if each variable is quantified consistently. The idea is that, at the start of the project, as you determine its overall value, it should equal 100 percent. As tradeoffs are made between each of the three variables, the project value will change. Together with your sponsor, determine what the threshold is for stopping your project!

The project tradeoff matrix documents the options available when critical tradeoffs must be made between these three parameters when weighed against overall value. This matrix captures stakeholder guidance as you control the project through execution. It should be considered your compass as you navigate the project through the sea of change. The highest priority parameter would have the fewest options for negotiation.

For example, from my experience, speed to market seems to be the key factor of success being defined by senior-level executives. Out of the three variables within the equation (scope, schedule, and cost), *schedule* seems to be the one most easily measured and demonstrable to organizational progress and improvement. The misperception is that by completing projects faster, organizations are able to be more responsive to market competition.

> Organizations commonly focus too much on speed to market. This misplaced focus can ultimately create a culture of just trying to hit a date instead of ensuring ultimate value is delivered. When this happens, you typically have a perceived notion that the organization is delivering. But in fact, the customer perception is that small, incremental, low value releases are occurring. This environment can allow competitors to steal market share. Stopping a project in this culture is extremely difficult. However, forcing a conversation about value may give you that opportunity to stop the project if its value has been degraded.

Using the Project Tradeoff Matrix in this example would show that "schedule" would have the fewest options for tradeoffs and negotiations. If the team is really going to meet management expectations and defined success, then discussions should really be about adding more resources (cost) or reducing the scope (scope to fit) so that the timelines can be met. An example of a tradeoff matrix is provided in Figure 1.3.

Figure 1.3 Project Tradeoff Matrix

	Fewest Options	Some Options	Most Available Options
SCOPE		✓	
SCHEDULE	✓		
COST ($)			✓

To be clear, just because you've completed this matrix doesn't mean you're done. Completing the matrix shows you the appropriate direction to take as you manage change throughout the project. As any change is brought up and assessed, you must determine whether you need to make any tradeoffs against these variables to determine the impact on the overall value of what your project is delivering. Your stakeholders will likely forget about the decisions they made about this matrix, so it's imperative that you keep it readily available so tradeoff conversations will be direct and clear.

Much like my recommendation for project success criteria, I highly recommend that your project tradeoff matrix is readily visible in your personal workspace, as well as in your sponsor's office. Project tradeoffs are never easy, because it means that something has changed and someone isn't getting something they want. These situations can be filled with passion and commonly come with a dash of finger pointing. This is when you need to direct the conversation, based on the original direction and success criterion that was given, to navigate your team to a clear path forward.

> Understanding the context of the change can help you learn a lesson now and determine whether other similar changes are more likely to occur in the future. It may be necessary to incorporate the change into the risk management plan and discuss how it might be mitigated in the future using the tradeoff matrix. Understanding this aspect of the change can help you move forward and help your project survive.

The project tradeoff matrix is a living artifact, meant to be updated and changed as needed. It's important to set those expectations with yourself and the team. For example, I recently came across a scenario where the project direction was originally set with the fewest options for cost (fixed budget) and the most options for scope.

This is the basic idea around scope to fit. While analyzing the product's features and functionality and comparing them with the cost to deliver the features, the team had essentially cut all of the extra possible features and functions. If the team cut any more items, the value to the customer wouldn't be realized and the project would no longer be viable. To bring the project's expectations and direction back to their original intent, the team had to revisit the tradeoff matrix. Now the matrix showed scope and cost having the fewest options. Schedule was the last lever this team could pull if it needed to.

The above example is important because it teaches the rule of "pick two" and when to apply that rule. I've listed guiding principles/rules around the tradeoff matrix to help you better understand how to use and update it. They're very basic but important to follow:

- Always start the project by picking only one constraint. Change will happen within your project and you will need options to balance along the way. Allow yourself and the team the ability to trade off where needed.
- Make sure that all decisions to change your constraints are recommended and understood by the entire team, but approved by the sponsor. The team can't work through this independently and make changes without communicating to the sponsor the reason for the change and the impact to the project. If you expect your sponsor to help make decisions throughout your project and to be positioned as the best advocate possible for the team and project, you must follow this step.
- As project change occurs, you are only allowed to pick two of the constraints (rule of two). Picking more than two isn't realistic and will only set the team up for failure. The team needs balance and the ability to adapt to change. Picking two constraints limits that but still allows it to happen. Picking three doesn't.

S.T.O.P. SURVIVAL TIPS

- **Stop** before you rush into any project. Take the time to understand the project's success criteria.
- **Think** about each success criterion and ensure that it's understandable, measurable, and achievable by the project team, stakeholders, and sponsor. Understand *who* is going to benefit from *what* aspects of the project and *how* they're going to measure them.
- The project tradeoff matrix will serve as your compass, directing you as you balance and manage your project against these success criteria. Remember the equation for determining whether your project has enough value:

  Value = Scope ÷ (Schedule + Cost)

- Remember also that the project tradeoff matrix is a living artifact, meant to be updated and changed when needed as changes in your project are assessed for overall impact on value.
- Don't hide your success criteria or tradeoff matrix! Once you've captured it, post it in the cubicles, lunch room, and offices of your team and your sponsor.
- Remember, to effectively lead a cross-functional team in delivering a product, it's vital that project managers have a crystal clear understanding of the purpose, scope, and success criteria of the project.

SECTION TWO

Team Performance

Knowing How to Build Your Team for Success

Stop and think! Can you do it all alone? No! Do you understand the overall strategic plan of your organization and how your project fits within that plan? Think about whom you need on the team to execute the project the right way. Each person and their skills will integrate and play a role in achieving the project's success criteria. Do you have the right team in place, both in terms of resource allocation and subject matter expertise, to successfully carry out the mission/project?

Observe your situation and surrounding environment. Understand the culture of the organization, your team, and your sponsor. What kinds of pressures or constraints are hanging over their heads that may influence how they behave and perform on the project? When project managers observe their teams in a purposeful way, they form better relationships and can become more responsive to their teams and sponsor.

Project management is usually about having all the accountability with very little authority. This will become very clear to you as you start to put your team together. You will soon discover that you'll get to choose very few, if any, of the team members. Typically, resources will get assigned to the project by functional managers. So how do you integrate individuals with varied knowledge and skills and organize them so that they trust and depend on each other to communicate openly and deliver quality results — working in a concerted way? Let me take a moment to illustrate some of the potential resources you may need on a typical technology project and the differing perspectives each one brings.

Product managers, marketing, and sales teams are typically enslaved to competitive features and customer requirements. They have one job and that's to ensure the customer is satisfied. On the other hand, product/system engineers focus their attention more inwardly, on standardized designs and repeatable platform components. Already we can see a difference in perspective on what some may think the drivers are of the project. Then user experience/product designers often gravitate toward usability and sexy aesthetics, whereas information architects focus on data flow efficiency. Each functional group seems to have its own master. And these contrasting perspectives on "what's most important" can produce individual behaviors that can be uncooperative and downright hostile at times.

I'd bet many project managers don't even see team construction as part of their job responsibilities. Differences in backgrounds, perspectives, cultures, and functional priorities are just some of the ingredients that can brew conflict within your project team. Consequently, it's critical for the project manager to proactively manage the dynamics and alignment of the project team and ultimately the team's overall performance. Project managers must structure their survival plan to take these divergent views and set them on a path for success. To create this structured plan, let's use the S.T.O.P. principles and again, tailor them toward project team construction and dynamics:

- **Social Bonding.** *Stop* and understand that social bonding should be built into your team with frequent communication, interaction, and group participation activities. Social bonding in a project team often will break down barriers, resolve conflicts, and increase the amount of information shared, thus further promoting integration.

- **Team Identification.** Allowing teams to be self-directed lets them create an identity for themselves, thus increasing individual motivation which can then increase their ability to cooperate. If a team is allowed the space to create their own identity — especially within the context of a larger organization — and the team's identity is shaped around shared objectives, this will increase their understanding of how they rely on each other and further foster cooperative behaviors. This autonomy can be more commonly referred to as "walling off" a team to ensure they're focused and allocated correctly to the mission. Additionally, empowering team members in this way shows them that their managers understand that they're able to work independently and achieve goals. That message alone can be very powerful in helping team members contribute to the project's success.

- **Orient Around the Work.** Project plans and schedules can't simply be put together using existing templates. Projects are not a series of checklists, managed with a "check the box" mentality. Team members and their associated tasks have to be oriented around the type of solution they're trying to deliver as well as the market they're trying to serve. A market-oriented project plan is a plan in which all team-member deliverables are recognized and related back to the customer's needs and requirements. By tying directly to customer use and value, team members become more sensitive and responsive to customer preferences and market conditions. Project plans based on overarching market requirements may direct project team members toward greater interaction as it becomes apparent to them that the common master they all have is the customer and ensuring they are satisfied. Teams focused on the customer realize that the solution they're building cannot be achieved by working alone.

- **Planning to Increase Performance.** It must be understood that formal planning has a positive influence on innovation and project activities. Product managers typically have the perspective that you can't plan innovation and that there's no role here for the project manager. They refer to the early stages of product development as "fuzzy," as so much is unknown relative to the ultimate solution that is finally delivered. To remove the fuzziness, more formal planning is needed, specifically in the areas of market research, competitive analysis, and requirements definition.

There's no reason why a tactical, formal plan can't be put in place to help manage these activities and ensure they're on the right path, getting the proper inputs necessary for the subsequent stages of the project. Formal planning during this stage is also viewed favorably — by the team

> and senior management — as disciplined behavior and thereby more credible than informal planning. Given higher credibility, individuals in a team are more likely to work together since the plan, in theory, is associated with a greater probability of success. Formal planning removes ambiguity in roles and responsibilities by clearly mapping out activities against expected outcomes relative to time. In contrast, when planning is informal or absent during this innovation stage, projects suffer from lack of clarity generating the perception that the front part is "taking too long" and will lead to compressed timelines.

You can see the importance of these four factors and their interrelationships. A project manager must take each one into consideration and construct a balanced approach to team construction. Now, let's look at how to use the S.T.O.P. principles in building a project team that will succeed.

Team Orientation and Integration Fundamentals

What differentiates those project managers who seem to have a successful team no matter what group of individuals they're working with — from those who seem to always have a dysfunctional team and troubled projects? The answer to this question was never very clear to me early in my career. I knew it couldn't be pure luck regarding who was assigned to their team that dictated their success. In my search for clarity, I read "Surfing the Edge of Chaos," an article by Richard T. Pascale from the book *Strategic Thinking for the Next Economy.*[4] In this article, Pascale wrote about how teams act like "complex

[4] Cusumano, M, A. and Markides, C. C. (Eds.) *Strategic Thinking for the Next Economy*. San Francisco, CA: Jossey-Bass, 2001.

adaptive systems." He noted, "What makes a system complex and adaptive is the ability to anticipate and be proactive." Said another way, human beings are complex and adaptive, and it's our ability to anticipate and be proactive that allows us to survive. Pascale helped me connect a project team's ability to survive and be successful with its ability to anticipate change and be proactive. Pascale also wrote, "Stable equilibrium equals death. For any system to survive, it must cultivate variety into its inner workings. If it fails to do so, it will fail to adapt to change successfully when it comes externally. While equilibrium endangers living systems, it often wears the disguise of an attribute. Species (people) are innately drawn toward stability and equilibrium — and the further they drift toward this destination, the less likely they're to adapt successfully when change is necessary."

We see this type of behavior on teams that have been working together for a long time. A once creative, highly functioning team at some point turns into a team that begins to bicker, finger point, and have items fall through the cracks because they have encountered a very difficult challenge. How do you cultivate a successful project team and prevent a once-functioning team from falling into the trap of complacency, unable to adapt to the new project?

The answer lies with something I like to call *creative disruption*. Creative disruption is the ability to periodically create a bit of disruption or change in the team to keep them fresh and functioning.

This change can be something as simple as co-locating team members or changing the development approach to the project. Implementing team change has got to be something that becomes natural and understood. Large, sweeping team changes never go well because the team isn't used to and hasn't recently experienced any type of change like this. The team has allowed itself to become stagnant, and big changes are perceived as completely disruptive. Small, steady streams of creative disruption in a project team can lead to better efficiency and productivity of its members and are better received than change handled in a big-bang approach.

Creative disruption can also provide continuous improvement throughout the project. Often, project managers wait until conducting a lessons-learned exercise to identify opportunities for improvement, and then they don't implement the lessons until the next project. By that time, the project is different, the team is different, and the lessons are old and may no longer apply. Stop waiting and start addressing! Continually evaluate the project and implement small changes to improve team efficiency and delivery *during* the project. Be creative if you have to. Creative disruption will have a very dramatic effect on building team bonds and furthering team integration as you strive to create that "ideal" team.

Creative disruption is a method to stop or disrupt the monotony of the everyday and introduce small, incremental changes that keep the team challenged and fresh. Just because something worked for a team on a prior project doesn't mean it will work on the next project. Keep the team on their toes and in a mode of constant, critical thinking. Your project results and team dynamics will dramatically improve as this technique prevents stagnation and complacency.

When we talk about what makes an "ideal" project team, it's important to understand what your ideal state is and work backward from that. Your ideal state should be based on an understanding of what it means to have the right people, involved at the right times, doing the right work, and acting in concert as a team throughout the development process. To identify your ideal team state, you have to use the S.T.O.P. principles above as, more or less, ingredients to a recipe — a recipe for team building.

The type of project will help determine how best to construct the team. For example, a new innovation-type project, where the product or solution doesn't currently exist in the market place, is likely to require more activities around social bonding, further thought and analysis on team identification and stratification, more planning of project communication and socialization, and more cross-functional interaction. In addition, more complex and extended launch activities are needed, considering it could take up to five years before a new product achieves the expected return on investment (ROI).

Every person has the basic, fundamental desire to succeed, make money, and provide for their family. Stopping to understand what drives a person professionally may help you determine what types of incremental changes will work to keep your team fresh and functioning. This analysis is another way to determine your team's DNA and how each component (person, function, project, etc.) fits in. Understanding the team's DNA can provide you the context you need to plot your path to project team integration and success.

On the other hand, an incremental development project, where only minor feature enhancements are being made to an existing product, is already understood by team members from a context and scope perspective. Less analysis is needed because there are fewer "new"-type tasks and activities. Product positioning and pre-existing sales channels already exist, the product technology is typically understood, and the rollout and support of the product probably has an existing infrastructure in place. Far less team bonding will be needed because bonds have likely already been established. The project work will be far better understood because it will already be familiar, so the team will know how to naturally orient around the work.

This isn't to say that this type of project is going to be a slam dunk in terms of delivery and success; but, it doesn't typically require as much thought around team construction and dynamics as the new innovation project in our first example. However, don't get caught into the complacency trap Pascale described. Because the project feels so familiar to the team, they could get too comfortable with the effort and begin to get "sloppy." The aforementioned incremental product enhancement-type projects are the perfect candidates for creative disruption.

Understanding the differences in characteristics within these different types of projects will give you the framework you need to construct your team and build communication and stakeholder management plans around it. Figure 2.1 further delineates the types of technology projects you may encounter and associates them to certain project characteristics that will help guide you in determining your next steps for team construction.

Figure 2.1 Project Types

PROJECT TYPE	DESCRIPTION	CHARACTERISTICS
New Product Development/ Growth Initiatives	▪ Developing a new product or service (also includes new platforms for delivery) ▪ Products that will produce new revenue by selling to current customer base and/or an entirely new customer base (market expansion opportunity) ▪ New product development may be done to develop an item to compete with a particular product/service or to improve an already established product.	▪ High rate of return/high risk ▪ Most likely to be killed off earlier ▪ Typically multi-year investments and paybacks ▪ New technologies and business problems being solved ▪ Extremely high senior management communication and oversight ▪ Typically, team members have never worked with each other, so social bonding activities are highly recommended. ▪ Organizational capabilities may not exist or be mature enough to deliver the desired solution.
Incremental/ Sustaining Product Development	Project investments in this category create enhancements to existing products, platforms, or other infrastructure that are done to protect or expand existing revenue streams.	▪ Typically less risky and less likely to be stopped ▪ Usually last 6–9 months ▪ Existing technologies and business problems are understood. ▪ Senior management communication and oversight is high but not as reactionary as in new product development. ▪ Team members likely to have worked with each other before. Social bonding activities are not as necessary as with new product development. However, creative disruption techniques may be required to keep the team fresh. ▪ Product launches are normally less resource intensive and generally quicker returns on investment are received.
Maintenance	Projects and product investments that define the minimum annual operating cost to keep products functioning in an "as is" state and to provide the support that customers are entitled to.	▪ Low risk ▪ Usually last 3–6 months ▪ Existing technologies and business problems are understood. ▪ Senior management communication is more on a cadence structure. ▪ Team members have worked with each other extensively.

Start by *thinking* about your team and the types of roles and role interaction you'll need to create success.

To do this, we'll need to start at the macro level of project and product development characteristics and work through the fundamentals step by step. To construct your ideal state, you must break the roles of your cross-functional teams into three categories: product management, product development (business analysis, engineering, quality assurance (QA), etc.), and manufacturing/operational support. You need to do this deconstruction because these are the fundamental roles that will need to interact with each other, through varying degrees, throughout the project. You also must simplify the product development process to three categories, which I call the "three C's" of product development: conceive, construct, and commercialize. These categories are also fundamental to considering the types of activities that need to occur during a product development project. A high-level diagram of this process is provided in Figure 2.2 to help associate and frame the next section.

> Understanding team member roles and responsibilities is another way to force you to stop before acting. Don't assume just because you have the right job titles on your team that you have the right team. Make sure you and everyone on your team understands their roles and what everyone is counting on them to produce. This step is foundational to your project success. If your team members don't understand how to rely and depend on each other, your project will be riddled with confusion, duplication in work, and gaps in assignments — with all three happening at the same time.

Figure 2.2 Product Development Process

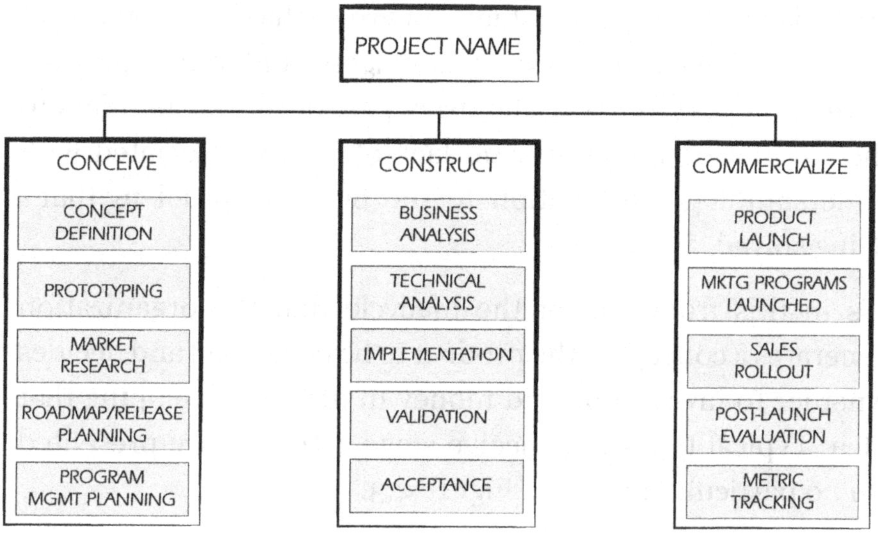

Realize that having everyone involved all the time isn't necessarily the answer. A high level of cross-functional integration may not lead to more effective and efficient results under all circumstances. Many times we, as project managers, get right into the weeds of trying to force fit a team and the project work into some predefined methodology. To avoid this, I recommend using the S.T.O.P. principles and starting at the macro level when developing your project plan. *Stop* and get the fundamentals right first, then refine from there.

Let's talk about each one of the product development stages (conceive, construct, and commercialize) and how best to structure team integration in each one, based on where you are in the process.

Conceive Stage — Removing the "Fuzziness"

Historically, the conceive stage of a project has been somewhat of a "black box" to project managers and other product development resources. The fuzzy part of this stage refers to the amount of ambiguity and lack of clarity associated with understanding the solution to the business problem that's being solved.

It's in this front end of the lifecycle that the organization generates a concept of the product to be developed and decides whether to invest time and money in further developing that idea. Typical types of conceive stage activities are presented (in no particular order) in Figure 2.3.

> There is an education component that many project managers find themselves responsible for. They have to ensure that the sponsor (who could be the vice president of marketing) and others understand the role of the project manager. A lack of understanding may cause some sponsors to inadvertently restrict the role. This is why you must be a proactive, vocal project manager. If project management is to become the true value-add discipline to the organization, then project managers must become champions of the overall effort and not just passive task masters.

Figure 2.3 Conceive Stage

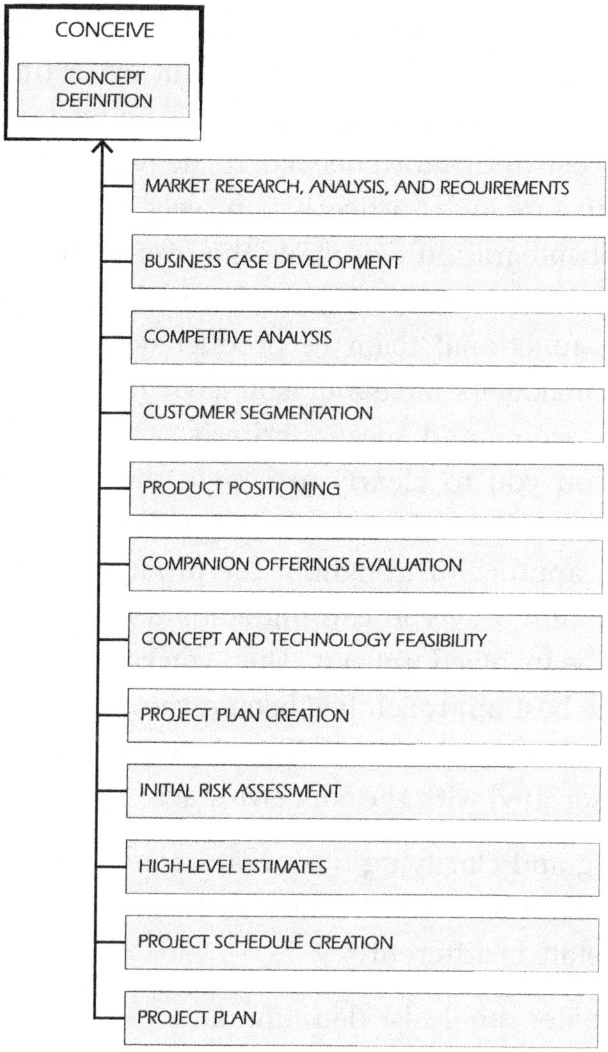

Project and program management is typically completely underutilized during this initial phase. Because of this underutilization, this part of the project is often not included when planning and tracking development cycle time. This is a big mistake that will hinder a true understanding of the overall cost and effort needed to deliver the solution. However, if applied properly, project and program management integration can make the overall development process more effective and efficient. Remember, to effectively lead a cross-functional team to product delivery, it's vital that project managers have a crystal clear understanding of the purpose, scope, and success criteria of the project. The team relies on you to clearly articulate the boundaries of scope, understanding of the budget, timelines for delivery, and delivery approach throughout the product development process. The only way you can understand the needs of the project is to be involved upfront. Then you can build the plan to ensure the best approach has been taken. Specifically, the following tasks should be considered project management activities associated with the conceive stage:

- Defining and clarifying business drivers for the project describing the current state challenges and how the future state is different.

- Stakeholder analysis defining and clarifying project success criteria. We mentioned this earlier in Section 1, but specifically, this is when this activity needs to occur relative to the overall lifecycle.

- Core team and extended team identification. It may be early in the project, but establishing the core team prior to starting project activities is critical. You don't want product management running ahead with context and market requirements without the rest of the core team involved and benefitting from their work. Extended team members should also be identified at this stage to determine how and when they should be brought into the project.

- High-level risk assessment focusing on external risks. External risks are risks related to the external market or customer base. For example, is there a competitive risk of someone else developing this solution sooner than you can? Have you confirmed that there's a willingness to buy this product and a market that supports this solution, or is there a risk that no one will buy it and thus, no associated ROI? Questions like these need to be asked to assess the external market risk to your project, to further validate whether it's viable.
- Creating a project schedule to include market research, competitive analysis, customer interviews, business case creation, and technology assessments.

This is where you remove the fuzziness from the front end of the project and understand all the steps necessary for the team members to complete this type of work. Your challenge will probably be that these types of resources (product management) typically aren't used to working with a project manager this early in the process. Don't be shy; introduce yourself and help them put a plan together that makes sense. This is where your analytical skills will be critical: ask a lot of questions to ensure they understand the work and effort required for their deliverable before they start working on it or begin setting unrealistic expectations and deadlines. This step is also extremely critical in understanding resource allocations and team assembly. The sooner you can understand this effort, the quicker you can put the right members of the team together to help forecast additional work later.

The other main task of the project manager during this stage is to ensure other team members have the information needed to move forward in the development process. Typically, your plan will only be as good as the level of understanding acquired by your team. Increased integration of product management and product development resources during the conceive stage increases context, awareness, and efficiency in the overall product development process. Product management-type resources can help provide perspective and explain the voice of the customer to resources like engineers, business analysts, and system architects.

With this upfront involvement, these product development resources can ask critical questions about the problem statements, brainstorm potential solutions, and ultimately get back in front of the customer with concepts to vet sooner, before true coding and development has started. In addition, the more project managers, business analysts, system engineers, architects, etc., understand upfront, the more likely they are to uncover landmines and pitfalls earlier in the process. Finally, the upfront cross-communication between these two groups helps with estimates for the rest of the project. Both groups have a better chance to truly understand the project constraints, impacts, and each other's perspectives. This perspective and understanding is extremely important in avoiding estimates that appear out of sync with what is being asked, which could lead to lack of confidence and trust within the team. The positive outcome from this upfront communication is a shared commitment and/or shared accountability for the project and the work that lies ahead.

Integrating other types of resources like manufacturing and operational support staff during the conceive stage, however, doesn't really prove valuable or seem to have any direct correlation to increased efficiency or effectiveness. Typically, the conversation at this early stage in the process isn't relevant enough for those types of downstream resources and doesn't provide them with any further clarity on what actions they should be taking regarding the project. Including them on the periphery — as an extended team member — and keeping them informed of the progress at this early stage is preferred and a more efficient use of their time.

Just to be devil's advocate, when dealing with projects where the purpose is to improve operational efficiencies or where the customer of the project is an internal operations- or product support-type resource, the role of the operations function changes. In this scenario, the role of operations staff moves from support into more of a product management-type role where they are defining requirements. In this scenario, stop and ensure that everyone understands their role and responsibilities. Revisit the role of the product manager and assign this to the operations staff member who will be accountable for the solution and associated requirements.

It's this stage when projects are most likely to be stopped, especially new product development-type projects. Be smart with how you integrate resources at this early stage so that team utilization is efficient both in cost and time. There is no need to spend project time and dollars if the project will be stopped. The conceive stage ends when an organization approves, funds, and begins formal development of the concept — thus when things become clearer and less fuzzy.

Construct Stage

As we move further down the process, the construct stage is when you need to make sure all functions are tightly integrated. Early in this stage is when marketing, sales, and manufacturing plans are created. Later in this stage is when channel development and launch/rollout plans are developed in addition to the typical deliverables associated with product development (i.e., functional specifications, technical specifications, quality assurance plans, validation and verification plans, etc.). Figure 2.4 is a high-level diagram of what the construct stage may look like within your organization. All parties need to have a clear understanding of both the high-level project schedule and any additional details on feature/function value, technical design and support considerations, platform considerations, pricing and contract structure, etc., to better support the overall initiative and ensure that all deliverables are in place for a successful project completion.

Figure 2.4 Construct Stage

> Using the project's business case as a mechanism for stopping a project is crucial. Make sure the team realizes the importance of the business case and its dynamic nature: it is a living document. Don't let the business case sit on the shelf and collect dust. It's fundamental in understanding the value of the work being performed and needs to be re-assessed any time changes occur in your project.

Be sure that your team hasn't forgotten what's driving the project by going back and revisiting the project success criteria and business case. Ensure the business case is kept up to date. Although the business case is typically developed and owned by the product manager, any expectations and long-term vision it contains must be understood by the entire core team. From my experience, even good change control measures during this stage typically leave off this critical step — ensuring that any changes fail to trickle back up to the business case. Project change is fine, but if changes lead to dramatic alterations to the business case, then the business case may no longer prove true and you should pursue other discussions regarding the viability of the project.

Integration between product management and product development is critical during the construct phase. Communication should be constant; and to be sure it happens, communication sync points should be called out in your project schedule. Don't assume one group is talking to another. Assumptions in communications and cross-team interaction usually lead to lack of clarity, confusion, and potentially the wrong solution being developed downstream. Confusion within these two groups can't be allowed.

Agile management practices, which include scrum, were originally conceived around fostering and improving cross-functional communication. I have found that many of these practices are excellent when dealing with highly collaborative environments. The scrum reveal method is one way to increase the transparency and trust within your project and ensure communication is constant. Reveals allow developers and other team members to showcase the work they're putting into the project. Skilled people typically are proud of their work; allowing them to showcase it will give them a better sense of purpose and pride toward the overall project. Reveals also provide other team members insight into understanding what is being built. This insight increases the opportunity for verbal feedback, early in the process — thus building trust across team members and furthering the collaborative environment.

My advice is not to wait until development starts before implementing these reveals. Reveals aren't meant to be just meetings run by engineers. In my opinion, reveals should be introduced in the conceive stage, when you're constructing market requirements and ideation occurs. In this early stage, product management is the group driving the reveal meetings and sharing context behind the concepts and requirements, as the idea is being fleshed out. Reveals at this stage also provide you and the team the ability to collaborate earlier on the design concepts — an approach that will further reduce the overall product development time. The quicker your team can start meeting and communicating like this, the sooner you can demystify the work associated with the project and the better chance you have of developing stronger relationships and trust across your team.

> Agile is intended to enforce team commitment as each member of the team is not only assigned work in a sprint, but also is committing to completing that work in that timeframe. This shared and public commitment is a key tenant in a team approach to Agile development.

Another communication concept adopted from Agile techniques (XP - Extreme Programming) is the 15-minute daily (or at least every other day) stand-up meeting. The stand-up meeting allows the team to walk through project status and discuss any roadblocks, issues, or potential risks about to be triggered. Don't let this type of meeting seem "heavy" to the team members. Be diligent about keeping the meeting to 15 minutes, and take any lengthy discussions off-line between the appropriate parties. Just ensure that the off-line conversation does, in fact, happen.

One method of ensuring product management and product development keep communicating is online documentation. What I mean by that is taking the construction of the deliverables out of the typical word processing software and moving it to a place where it can be viewed in near real-time as it's being created. This concept is extremely useful when new functionality is being proposed or larger projects are being undertaken. Documenting online shows complete transparency among the team and allows clarity to be obtained sooner rather than later. Again, I'm a big fan of starting this process with product management while market requirements are constructed. Having insight into what's being requested and access to project progress in near real-time can get the team talking sooner about design options and solution creation.

Don't let design get out of control by waiting until the design is essentially complete before it's presented to product management for review. This is a path leading to rework and timeline slippage. Move these deliverables online to allow product management to review how the design is being thought about and constructed. Allow them a chance to provide feedback and clarity as soon as possible during the process. This method, coupled with the daily stand-up meetings and biweekly reveals, will ensure that the right conversations are happening around the right information, clearing up any confusion and issues along the way. These three methods allow the team to focus on communicating around information *in* the deliverables and not just around the completion *of* the deliverable – allowing you to build a plan that is relevant and appropriate to the work at hand and to keep your schedule up to date in near real-time.

> The concept of online documentation is just one approach for ensuring early review of deliverables. However, the point of this example is to drive home the importance of including early reviews, drafts, reveals, etc. into the overall process of the project plan and schedule. By building it into the process, you're inserting additional quality assurance and allowing reviews and improvements to happen continuously and frequently, preventing error identification from happening too late.

During the construct phase, integrating product development and manufacturing operations resources shows a strong positive impact on efficiency in the overall process as well.

The sooner manufacturing understands the details of the design and materials, the better they can prepare for efficient production processing. Product development may also consider information about restrictions of existing production capabilities and the impact of desired quality levels during development so that costly modifications later are prevented.

Integrating product management and manufacturing/ operations can also increase project efficiency and effectiveness during this stage. Production processes and the product market introduction are prepared during this time, which involves coordinating production schedules and sales plans and therefore requires higher amounts of and more frequent information exchanges between these two groups. Again, as a project manager, you must ensure that the right people are talking about the right deliverables at the right times in your plan. The construct stage ends when all product development work has been completed and your project is ready to launch.

> One of the biggest challenges for project managers isn't managing the engineering and QA staff, but understanding that this same type of discipline and management is needed for other parts of the organization as well. Project managers need to stop just managing parts of the project and start recognizing that it's their job to oversee the entire project. Those extended team members also have key deliverables, and the project can't be successful without their engagement, input, and execution.

Commercialize Stage

Although the product management community believes the conceive stage is the only part that is fuzzy, now the launch and commercialization activities have also become fuzzy. Some of the reasons for this increased fuzziness are the further complexities involved with true solution delivery and market-facing product development. The project manager has become increasingly focused on development, neglecting both the beginning stage when the solution is defined and the end stage when the solution is being produced and commercialized. Project managers seem to be "celebrating" too early — calling for the end of their projects when coding is complete instead of when the successful product or project result is in the customers' hands. This premature celebration is usually followed by a fire-drill, where all of the planning staff and individuals associated with product launch-type tasks are suddenly pulled together to figure out how to implement customer requests, address marketing and launch events, and support the product after it's already in production.

As you near completion of the project and approach the commercialize stage, as shown in Figure 2.5, preparation and realization of market introduction needs to be coordinated with a focus on effectiveness. Typically, from a product development standpoint, most, if not all, of the development work has been completed and approved by this stage.

Figure 2.5 Commercialize Stage

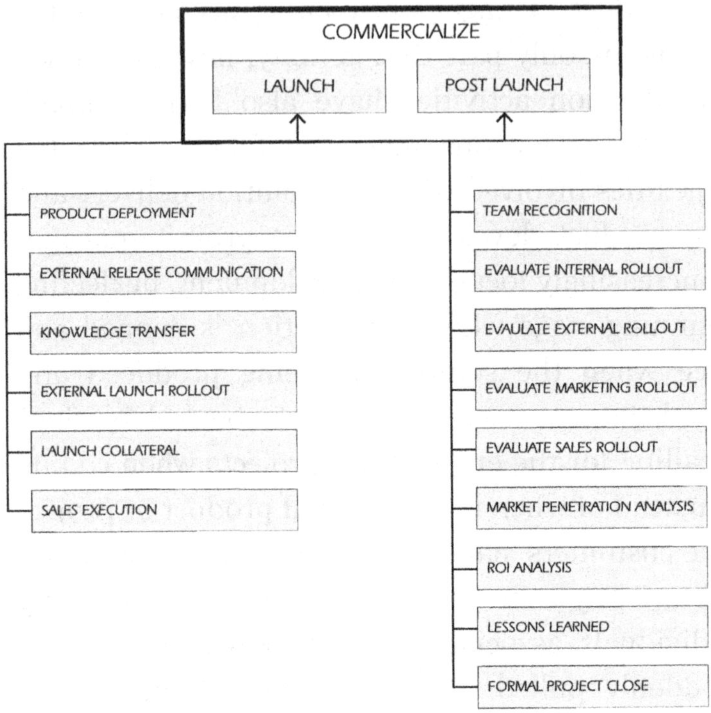

Product managers and marketers are usually very focused on customer discussions, sales training, event planning, media events, etc., where product development resources aren't needed to execute these types of activities. There aren't too many handoffs between these groups during this stage, so increased integration probably won't change the outcome or velocity of the project. However, product development needs to work with product management to understand when the product is going to be available for launch and what the overall strategy is around its rollout. Product development needs this information to ensure that adequate resource planning and support is available through the execution of pilot customers,

The question of "to beta or not to beta" seems to come up, all too often, late in the project. The use of a beta as an approach for market introduction could be key in stopping the effort of a full-blown product launch if additional validation on the value proposition is needed. Ask this question upfront during conceive and NOT during commercialize. Waiting too long to ask this question could throw your entire team off target and cause you to stop and re-plan. And in this case, stopping will not be perceived as favorable.

beta releases, initial internal and external questions and feedback, issue escalation and resolution, etc.

As long as the rollout plan was clear, product development should have enough insight to allocate the necessary resources to support the product. Again, keeping all groups informed of each other's activities and progress is preferred, but tight integration may not be required.

Integrating product development and manufacturing is even more important during this phase with regard to efficiency gains to the project. Preparing for a smooth production start requires frequent interaction and information exchanges between both groups.

The major risk of loose or inadequate integration during this stage is the organization being unprepared to deploy or launch the product. Organizational and market readiness is critical to the project's success. If you haven't prepared your organization for the launch of the product, you haven't done your job. Make sure you've verified that all documents and deliverables created by extended team members and business representatives are complete and clearly understood by all team members required to support your launch date. Figure 2.6 provides a summary to help visualize the relationships between each of these groups relative to the product lifecycle.

Figure 2.6 Cross-Functional Integration Across the Product Lifecycle

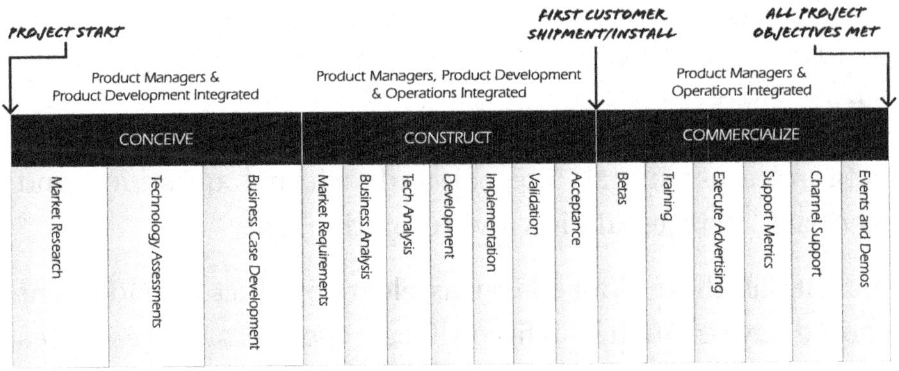

I once witnessed a project where the product manager constantly prodded the development team to complete the project on time. This nagging caused the project manager to focus internally on the development team to make sure their schedules were tight and progress was well communicated. The project suffered from constant changes in scope from product management changing requirements as well as an unusual volume of staff turnover due to organizational changes.

During this time of rampant change within the project, the project manager did an outstanding job adjusting the plan, communicating all risks, implementing a change control process, and informing all stakeholders of the project's progress along the way. Again, the project manager's focus was on development and the internal progress of the project.

In the end, the development team ran across the finish line and completed their work on time, within scope, and on budget, only to find that the rest of the organization wasn't ready.

The project manager constantly communicated the state of the project and tried to prepare all extended and supporting team members for the launch date. However, when it came time to launch, both product management and the operations team weren't ready, and the project sat on the shelf for six months while those functions took their time completing their deliverables. This happened because as the product management and operations deliverables started to show delays in delivery, the project manager completely overextended himself and tried to complete the tasks of those functions — single-handedly trying to carry the entire project on his shoulders and into the hands of the customers. Because the project manager took his eye off the overall big picture of product delivery and only focused his attention on the development part of the project, he found himself taking on more accountability as he tried to rectify the problem.

By this time, it was too late for heroics. Instead of integrating product management and operations earlier during the construct stage of the project and forcing communication and accountability, the project was late, allowing a competitor to catch up and launch their own version of the functionality one month sooner. Sure the development and the engineering staff finished on time, but the overall outcome of the project was failure.

Understanding the success criteria of the project and how to break down the product development lifecycle to ensure that optimal resource integration is occurring at the right times is fundamental to constructing your project survival plan. No one ever remembers the heroes of the project. All they remember are the outcomes. If the outcomes miss expectations, then the project has failed its mission. And trust me when I say you never want to be remembered for a failed project.

Project Stratification

It's important to understand the amount of governance required for your project and whether different team stratification layers are needed. The complexity of your organization and how many teams need to be involved will affect your communication plan.

When I talk about team stratification, I'm specifically referring to how you structure different members of your cross-functional team so that their interaction with the project is relevant and productive, with the appropriate level and frequency of involvement. Communicating everything to

> Yes, the majority of project management revolves around communication. However, too much communication and too much governance can dilute your message and drag down your team. Reformatting the same information over and over will cause your team to stop critically thinking about what they're doing and what they're intending to say. Communication will start to get sloppy and unproductive. *Stop* and *think* about the governance structure of your project, ensuring the right mix exists with enough uniqueness for the message's customers, thus preventing over stratification and over communication.

everyone isn't necessarily the answer. Too much information at the wrong time could actually have an adverse effect on your project and cause some team members to disengage and become apathetic. Too much information to the wrong people can also make your life as a project manager a living hell — you'll have to answer every question and explain every bit of detail you just tried to communicate through status updates. Keep in mind, I'm not recommending you don't communicate with certain cross-functional team members. What I'm recommending is that you structure your communication plan appropriately so that you're communicating the right level of information to the right people at the right times to allow their interaction with the project to be efficient and to ensure their interaction doesn't interfere with the other team members' progress. Be thoughtful and concise with your language and be cautious of incomplete thought, which could trigger more questions by those that don't understand and waste more of your time providing explanations.

Figure 2.7 describes the different types of team stratification that you may need for your project. Understanding what type of project you have and how each team may influence the overall outcome should help you determine how stratified your teams need to be. I'm not suggesting that every project needs to have all the support teams or governance implied by the number of teams below. You'll need to map out the influence each of the teams has on the project and determine if they are important or critical to success. Fundamentally, I believe that every project needs a small "core team" of individual subject matter experts that essentially become the cross-functional leadership for the project. Regarding true cross-functional integration, the core team is my one real requirement. However, the table is meant to give direction and provide rules of thumb if you do intend to use any of these team concepts.

Figure 2.7 Team Structure Stratification

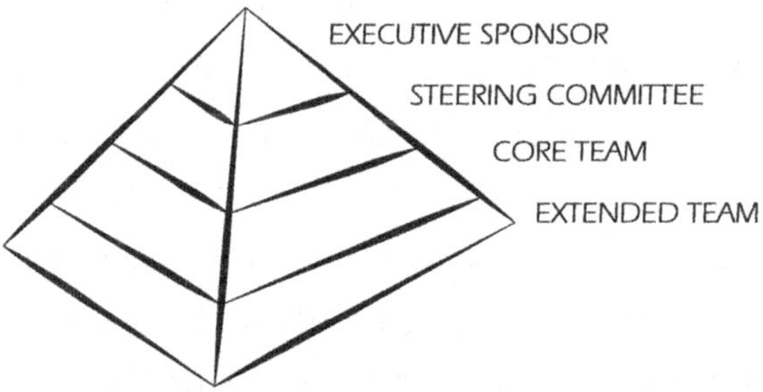

TEAM	CHARACTERISTICS
Executive Sponsor	Ideally one executive sponsor, who should be both financially and emotionally committed to the project. Needs to have the authority to continue the project as well as stop it, if necessary.
Steering Committee	Typically made up of key stakeholders and resource controllers. This group often seems like a necessary evil; but if structured correctly, it can be effective and ensure appropriate project governance. Role should be to allow the team to navigate an organization's matrixed environment and remove roadblocks. Roadblock removal will actually be a majority of the work that this team does, as opposed to actual "steering."
Core Team	The project's brain trust. Small group of 4–8 people who dedicate at least 75 percent of their time to the effort. Most of the time, it will require a lot more. Meets weekly to discuss the project; the idea is to gain consensus within the core team and not the entire organization.
Extended Team	A much broader group of individuals. These are key contributors who will need to be brought in and out of work and the effective communication of the project as the initiative proceeds.

Establishing the Core Team

As mentioned before, it should be understood that as a project manager, you won't be able to hand pick your team. However, what you *can* do is make sure your resource requirements are well understood so that as functional managers begin assigning resources, they assign resources based on project *needs* instead of simply on who's available. Your initial goal should be to establish a core team that has experience and a proven track record for being transparent and successful.

Core teams are made up of cross-functional team leads or representatives. Core teams complete a set of interdependent tasks within each phase of the project to produce the deliverables necessary to complete the task. Each core team is made up of approximately four to eight members, depending on the type of project and stage of the project relative to the product and project lifecycle. These roles can be filled by a number of different resource types at the discretion of the functional managers and the project teams. The team members that typically could be found within the core team are:

- Product Manager/Business Lead
- Project Manager
- Business Analyst/Systems Engineer
- Technical/Engineering Lead
- Quality Assurance Lead
- Manufacturing/Operational Support Lead
- User Design Lead

It's important that each member of the core team understands that they are accountable to each other, the business, its customers, and the executive owners of the project. The core team should be considered "family," sharing a commitment to the project and open communication and where an inner circle of trust, transparency and reliance exist.Brainstorming, problem solving, and schedule construction sessions, as well as external social events, are the types of social bonding activities needed to allow these characteristics to flourish among your core team and within your project. As a project manager, you should be building these types of activities into your schedule. In most cases, you should be facilitating the sessions as well. Be the glue that binds the team, and allow them to gel around you. By facilitating team building, working, and bonding sessions, you can create an interactive and engaging environment instead of one dominated by one person on the team. Figure 2.8 represents this core team relationship.

> Keep in mind that this view of the world is from the project perspective. Once the project to build the product is complete, the project manager frequently moves on but the product manager stays with the product through the entire lifecycle. Therefore, the product perspective would have the product manager at the center of the core.

Figure 2.8 Project Core Team Relationships to Project Management

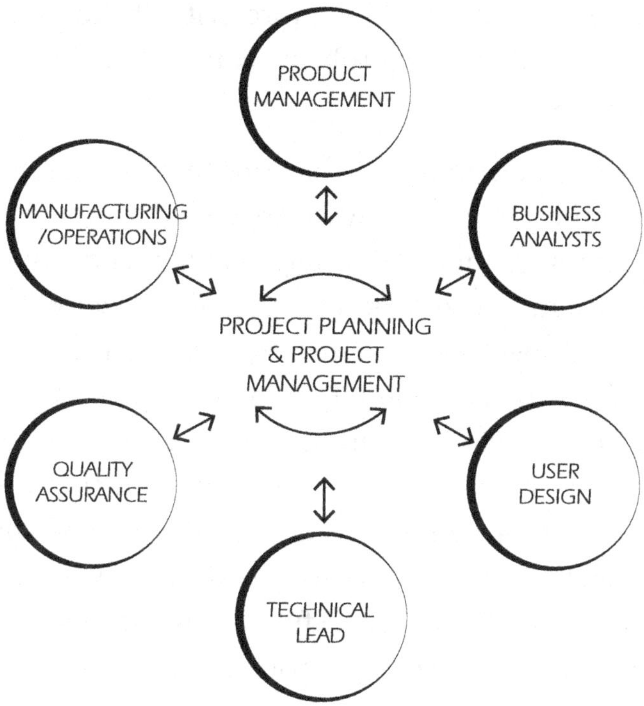

Also look for core team bonding opportunities outside of the schedule. Getting coffee in the morning, going out to lunch as a team, and attending company functions and corporate meetings as a team are all ways to strengthen the bond. As a project manager, be the one responsible for making sure this group dynamic happens. Be the leader for this team both inside and outside of the project.

In my experience, core teams make better review and project decisions than individual team members. I believe this team concept is especially true for projects where new products or product features are being considered and the design, commercialization, and marketing approaches are virtually unknown. Each member's unique perspective should be heard in order to help assess options. By using this team approach, you allow more critical thinking, which typically results in a more robust, comprehensive decision-making process.

It's also the responsibility of the core team to communicate efficiently and effectively with extended team members. Extended team members typically include sales, marketing, implementations, user access, legal, accounting, and compliance. Project teams must regularly engage extended team members to keep them up to date on project status and deliverables. I typically recommend that, at a minimum, core team members meet twice a week to stay on the same page and help each other remove obstacles and barriers to progress. This frequency may increase to 15-minute, every-other-day stand-up meetings as you move further into development and clear communication becomes more vital.

I then recommend meeting with extended team members bi-weekly, to keep them up to date on project progress and discuss any upcoming project deliverables they may have. It's important to continually communicate the project status and constantly remind the extended team of the project goals, success criteria, and constraint priorities (scope, schedule, and cost) under which the project is operating. This approach will drum into the heads of everyone involved what is driving the project and the basis for how the team is making decisions.

I'll provide a general description for four of the eight possible core team roles above: product manager, project manager, technical owner, and business analyst. I believe that these four roles, at a *minimum*, are critical in establishing a core team that's highly functional at both the project and product level. My reasoning around this "critical four" comes from an insight sparked by one of my previous managers. He referred to these four roles as the "kite" of the project. He would draw a kite on his whiteboard and ask "who is filling each one of these roles?" If we could answer that, his next question would be "who's acting as the tail and the ribbons on the tail, from an extended team perspective?" The idea was that these were the people who were going to be critical to the project if this project (kite) was going to fly. If we could answer these questions in 60 seconds or fewer, then we had a pretty good idea who the key members of our team were and the communication that was needed between them. Figure 2.9 illustrates this kite relationship.

Figure 2.9 Key Project Core Team (Kite) Structure

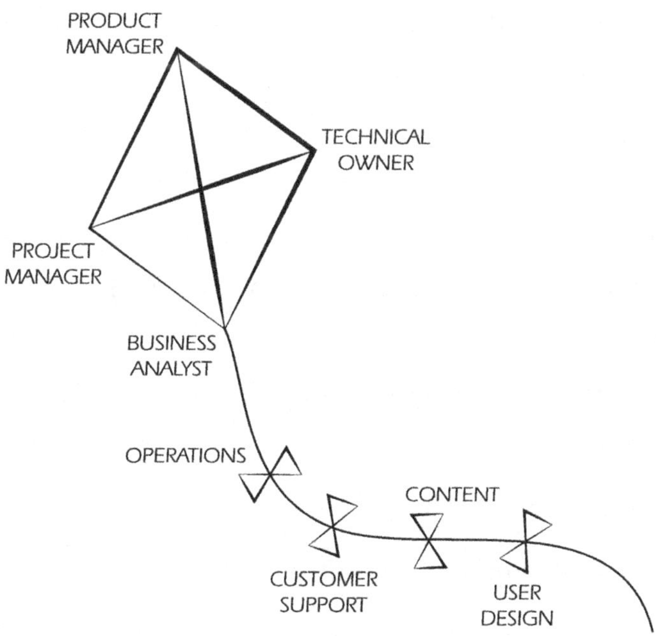

As I further describe the responsibilities associated with these four key roles, I'll pay particular attention to the project/program manager role because this is the focus of this book and my area of expertise. It was important to include the other roles for context and understanding of how the project manager's responsibilities fit in with the other core team responsibilities. However, those other roles and responsibilities may be different within your organization and you should use this description only as a guide to make sure someone within your core team has that work covered.

Product Manager/Business Leader

The product manager can sometimes be described as the market champion responsible for interpreting the voice of the customer and being the functional owner of the solution. They're the ones the organization holds accountable for achieving business results (as opposed to project results). In iterative-/incremental-orientated organizations, this person's title/role might be "product owner."

Typically during the conceive stage, the product manager performs the following activities:

- Partners with finance in the value determination of the project (e.g., identifying the expected ROI and impact to the market)
- Owns the contextual understanding of the problem statements
- Conducts market research
- Performs competitive analysis.

- Develops product vision (three-year plan) along with associated rolling 12-month tactical product roadmap
- Manages overall product investment funding relative to the business objectives and goals. This could mean balancing between new, growth initiatives and maintenance-type fixes and upgrades; in other words, they determine priority of scope to be delivered by the team.

As you move further into construct, the product manager's role further progresses into the one associated with providing input into the market requirements associated with the solution. They review and approve all functional, non-functional, and content requirements associated with the solution and the project. And they establish the "customer" acceptance criteria for the project as well as coordinate access and interaction with end-customers during development. Finally, during this stage, the product manager is vital to reviewing the test strategy and signing off on test cases after any QA walkthrough.

> The product manager focuses on observed customer needs and dynamic market conditions. This may result in product managers revising their requirements or desiring to add features, thus increasing the scope of a project. The project manager may perceive this as scope creep and want to keep scope as small as possible so that the project is delivered on time and under budget. This conflict typically exists within projects.

During commercialize, the product manager must ensure that the product launch is

> Good product managers know that all the features in the world will not matter if the project never makes it to market or if it's too costly and doesn't have an attractive ROI.

successful and that all supporting collateral and signoff has occurred. In addition, they own the prioritization of defects reported by the field, as well as any user acceptance testing needed for any defect. They may delegate these responsibilities to someone in QA or business analysis, but they own the overall outcome. Also, during this stage, they fundamentally own the decision and prioritization of defects relative to feature enhancements and potential new features in upcoming releases and projects. They are accountable for the success or failure of the product, extension, or enhancement with the executive team.

Project/Program Manager

The project manager is the one role that is accountable for managing the scope, schedule, quality, and cost of the project to a successful completion as determined by the executive sponsor. During conceive, the project manager must facilitate and consolidate project success criteria information and be sure it's completely understood by all team members and executive sponsors. They then create and establish the project plan (survival plan) document. This includes understanding the context behind the project as well as the associated budget and conceptual scope. Creating a high-level schedule with deliverables, resource assignments, estimates, and dependencies based on core team members and overall project success criteria also occurs during this time. This schedule can exist at the project level and be rolled up to the program level to show milestone dependencies between projects.

> Most discussions in the conceive stage are around ideas and inputs (pre-requirements) into the product production process. To remove the fuzziness and lack of clarity that can occur during this stage, more formal planning is needed by the project manager, specifically in the areas of market research, competitive analysis, and requirements definition.

As the project manager moves into construct, they're the objective person on the project responsible for providing early warning systems to ensure the team's success. They're typically also the person who must accomplish the project objectives and balance multiple projects at once to achieve the overall goals of the business. Without a doubt, they're accountable for managing the project, including:

- Driving the direction of the project based off of clear project success criteria, as measured through the tracking of project milestones and tasks
- Balancing competing demands for scope, time, cost, and quality — sometimes referred to as the fourth dimension of the aforementioned "iron triangle"
- Adapting the plan and approach to address the different concerns and expectations of the various stakeholders
- Driving the execution of the project plan
- Publishing management review meeting minutes to document meeting results and actions
- Assuring proper documentation of deliverables

- Maintaining the appropriate deliverables during product development
- Owning internal project team communications
- Managing the schedule with deliverables, resource assignments, estimates, and dependencies based on core team members and overall project success criteria. This schedule can exist at the project level and be rolled up at the program level.
- Confirming resource commitments (could be lagged depending on when projects are to be kicked off per the portfolio) to project artifacts to hit project dates. When they're lagged, the lag-time and any changes to it must be watched by the project manager.
- Instituting an "investment review" project status when resources/duration/external costs are seen as exceeding the initial threshold set by the project sponsor. This review could also be established as a review against multiple projects in one area where the budget has been set at the program level.
- Reporting, early on, projected changes to timing, scope, and resource load of remaining project activities
- Gaining formal commitment from all functions to execute the deployment plan
- Running the change management process, which is critical because scope changes always occur and the project manager is typically the moderator during the tradeoff discussion

> As you move through the construct stage, constantly ask yourself, what programs, plans, training or other events are needed for a successful launch? Keep assessing and refining through development but think beyond development... What does the operations function need to do to support the project launch? What does sales need to know to get the pipeline ready? Constantly ask yourself, "What do we do internally and with whom internally do I need to do communicate with while the solution is being developed?"

As the project manager moves into commercialize, they do NOT roll off the project immediately. They're accountable for managing the overall budget and schedule associated with bringing the product to customers. The project manager also ensures that production support activities and bug fixes are timely. They may even participate in or facilitate production support defect triage meetings. As you move to the program level, the project manager may be critical in reporting on program-level metrics (product-level defects, patches, support calls, customer satisfaction, rolled-up budget expenses, etc.). They may facilitate, consolidate, and manage cross-functional teams ensuring that extended teams aren't only prepared for upcoming project releases but are informed of any known defects or issues affecting the product.

Business Analysis/Systems Engineering

The business analyst typically works with the product manager to distill the market requirements into more specific functional, non-functional, and content requirements. They're accountable for understanding how the use cases are to be translated into actionable and implementable specifications.

> Business analysts are instrumental in driving requirements analysis during this stage. Requirements analysis is the process of understanding a requirement through the use of various techniques, including brainstorming, interviewing, and research, to solve a problem or problems. Deliverables from this process will be used by the project manager and the project team to develop estimates for the time, resources, and budget required to implement a solution or solutions that will fulfill the requirements.

Getting the business analyst involved during the conceive stage gives them more time to understand the inputs required to drive the solution. Their involvement during conceive should be planned for considering that the majority of their work happens within the transition between the conceive stage and the construct stage. By pairing business analysts with product managers at this early stage in the lifecycle of a product's development, organizations can optimize bandwidth, expertise, and interest related challenges that allow both roles to do what they do best — create value.

During construct, the busines analyst would be responsible for securing signoff on the functional and non-functional requirements by both the product management- and engineering-type roles. They would ensure that the market requirements could be traced to functional and non-functional requirements. They may also be involved in reviewing test strategy documents and signing off on test cases after a QA walkthrough.

Finally, during commercialize, the business analyst would

assist with defect analysis, write-ups, and specification updates where appropriate. They would also help product management prioritize and manage the defect backlog.

Technical/Engineering Owner

The technical owner typically determines the technical direction for the project. More than likely, the engineering team, as well as QA, may report to this role. During the conceive stage, it is critical that the technical owner understand the context of the scope of the project as it relates to the technical feasibility and capabilities of the proposed solution. Any platform decisions and high-level cost estimates for the project are typically decided by the technical owner during this stage.

As we move into the construct stage, the role of the technical owner becomes much more prominent because they are accountable for the technical vision, direction, and overall quality of the software/components. They typically work with the project manager to construct the technical aspects of the project schedule, identify dependencies between tasks, and assign engineering resources to perform the required tasks relative to their engineering discipline. Other types of activities associated with this role during construct are:

- Ensures that staff (including internal and external resources) are qualified to perform their jobs — including arranging for any additional training if needed
- Makes certain that the agreed upon software development process is followed
- Is responsible for the design integrity and quality of all

designs, ensuring that non-functional requirements such as product scalability or end-user response time are achieved

- Acts as the central point of contact for any hardware/software procurement
- Certifies that adequate technical reviews are performed and unit testing has occurred
- Guarantees that all software components are subjected to configuration management (where applicable)

As the technical owner moves into the commercialize stage, they're accountable for tiered support of the product. They need to ensure that the right amount of technical staff and the correct skill mix has been assigned to support the product. They're also responsible for technical issues that have been identified as "show stoppers" by clients and need immediate resolution.

Back to the Project Manager

The project manager should create a description of the roles associated with each core team member and document them in the project plan with associated deliverables. This association begins to give context to how the team is oriented relative to the type of work that must be accomplished within the project. It should be crystal clear who is accountable and responsible for each deliverable.

When designing the core team, the project manager should not only plan for typical development lifecycle activities, but take a broader system- or program-level approach. This program-level technique is crucial; it provides a proactive approach to staffing, allowing the team (and associated functions) to plan appropriately over the course of the *product's life* — not just the *project* life. Again, orienting the team around all the work necessary to be successful is vital. Drawing manufactured, artificial endpoints to your projects can be detrimental to your organization's and team's success. Typically, I find that a project's core team can execute and deliver well, but the organization falls down during support and commercialization activities because resources roll off the project and the organization assumes that zero resources need to be applied to support the product. *In other words, they underestimate that completed projects require a fixed amount of resources throughout a product's useful life.*

Another problem to watch for is *when* you establish the core team. As I mentioned earlier in this section, it's critical to do so at the very beginning of the conceive stage and NOT after requirements have been defined and design and development is supposed to start. If the core team isn't established at the beginning, then vital context, understanding, and guidance can be lost, leading to unnecessary challenges and conflicts downstream when the team finally does get assigned.

I want to be clear with demonstrating that you need to start thinking of your teams and how they're assembled beyond the initial scope of the project. If you're going to be a successful project manager, you have to fully understand how the project fits into the overall program or business objectives. Having a project manager roll off and leave the rest of the internal team to figure out how they're now going to manage and support the product can lead to a successful

> Stop prematurely rolling off of the project. Make sure the true endpoint of the project is understood based on when the success criteria and value is to be delivered. Rolling off too soon could cause a highly effective team and productive development process to be turned into a company fire drill and a market launch nightmare. Typically, that planning can't be done without the project manager being involved. I know you want to get to your next assignment, but it's not about the end date. It's about the value. Ensure that the value of your project has been realized before you roll off.

project, but at the expense of a failed product or overall failed customer experience with the product.

Organizations can increase the rate of return on their project management investment by assigning project managers to product domains for longer periods. Seasoning your project managers in the product domain results in greater continuity and less time bringing new members up to speed, thereby reducing costs to your business. It also allows the organization to utilize this valuable resource more during the early stages of the product lifecycle, as the product is newly introduced to the market. There's a lot of planning, managing, and process development and improvement that must occur during the early stages of a new product, after its initial launch into production.

A project manager is instrumental in ensuring that the product is stable and the resources supporting it are allocated appropriately and managed effectively. One could argue that the period when a new product is put into production until the time it is considered stable, is in fact a project in and of itself.

Project managers, business analysts, and technical managers need to learn to think longer term and be more strategic to ensure that organizations meet their overall business goals. Having said that, it's important for business leaders to realize that they need to make certain these resources *are engaged upstream in strategic planning* activities as well as downstream in tactical execution. These resources have skills that can be transferred from the project to the program and product. The better these resources are utilized across the entire product lifecycle, the better context and understanding will be achieved at the project level and the better the chances are for overall organizational alignment and project success.

The Core Team From a Different Perspective

Earlier in this section, we tailored our S.T.O.P. principles to project team integration and dynamics. We then talked about roles and responsibilities regarding different key functions and how they worked together as a core team throughout the different parts of the project lifecycle. I was straightforward, trying to keep the discussion very black and white. However, when talking about social bonding and team integration, it's important to look at your core team not from the functional standpoint but from the personality trait standpoint. This subject matter gets us into the softer skills needed by the project manager, because team dynamics mixed with personalities are never black and white.

In trying to manage the team dynamics of your project, it's important to understand the individual personalities that you may come across and how best to work with them. To that, I've come up with four basic core team personalities that seem to exist on every project. Now, I know there may be more, but the idea is to understand the basic four and overlay that on the two-dimensional core team kite structure described earlier to bring that necessary third-dimensional aspect to your project. The four basic types are:

- The Hero
- The Victim
- The Social Butterfly
- The Introvert

I'm no Freud and I'm not going to pretend to dive into the depths of psychological behaviors. However, you need to be aware that team dynamics are influenced by individual personalities and each team will behave differently relative to its individual team members. Let me define each type and explain how to best integrate them into your project team.

The Hero

The Hero on the project can sometimes be called the "cowboy." They're usually very charismatic, extremely talented, and hard to control. Typically, they're vague around deliverable dates and setting expectations, but they always seem very specific in pointing out issues that exist or that they've found. Remember, to be the Hero, they need to solve the impossible problem... a problem, nine times out of ten, that they've discovered. Now, I'm not saying the Hero is creating issues only so they can solve them... at least not consciously. However, because the Hero is so vague on commitment and detail, and because they are overconfident, they tend to overlook the obvious when performing technical analysis. This lack of analytical attention can cause issues later in the process. And this ultimately creates a scenario where they have to come swooping in at the last minute to solve the problems and save the day.

The way to manage the Hero is by micro-managing and pairing them with an analytical type during analysis. As a project manager, you have to mitigate the risk of unexpected issues being uncovered late in the process. To deal with this personality, you have to be the proverbial pain in their butt in terms of getting the detail around tasks, durations, and dependencies. All tasks need to last no longer than two days, with constant communication and follow-ups. If you are co-located, frequent, unplanned visits to their desk or office can help as well. To further mitigate the risk, you have to introduce them to their "new best friend" who is someone, typically from the QA organization, who can do that analytical thinking along the way and catch issues they might have missed. Initially, this forced relationship is a bit unwelcome, but if you diplomatically explain why you're pairing them, it can work out very well for both individuals. You have to tell them, "You are extremely talented and we want to make sure the most robust design is thought through upfront, so you'll have time to shine during its development. We also want to make sure we hit the date by having QA sit in and get a head start on their testing strategy. Ultimately, this should give us a superior test strategy and plan, as well as the best opportunity to succeed as a team."

It's not that you're feeding them a line; you're simply describing the relationship needed in a more digestible way. To get the most out of your resources, you have to harness their strengths and mitigate their weaknesses by complementing them with other team members. This approach fosters symbiotic relationships and social bonding on your project and prevents you from getting into situations where your Hero is more villian and your project needs last-second saving.

The Victim

The Victim on the project is usually someone who, like the Hero, is very talented and knowledgeable, but for some reason has a hard time delivering because of some sort of roadblock. This roadblock might take the shape of someone or something, but typically, the root cause, in their minds, is because they weren't involved or communicated with, and therefore, they're finding out necessary information too late, resulting in them having to push out their delivery date.

This personality is a bit harder to integrate into the team because, typically, there is some validity to their complaint. Initially, this leads you to seek out why no one communicated with the Victim. You'll have to ask, "How come you never included so and so? You know they're extremely knowledgeable and depend on that information." But the more you dig in, the more you'll start to realize that the reason the Victim wasn't involved was because they either didn't make themselves available or didn't agree with the direction, so they purposely didn't attend the meetings to discuss the project. They feel isolated and left out, but the fact is they created the situation. Others were forced to leave them out and move on just so they could continue to make progress.

So how do you manage this personality? First, you have to identify whether the situation they're describing is actually occurring. Talk to all parties involved and try and get at the truth, which is usually some place in the middle. First speak to the Victim. Ask them, "If you knew these discussions where happening, why weren't you proactive in getting the information you needed? Why did you wait for someone else when you didn't need to?" They have to understand that a part of their job is to be proactive and responsive and that they're still responsible for their work, so they better do everything in their power to get it done. At the end of the day, all the excuses go away, and the result will be determined by asking the question, "Did they deliver or not?" The sooner they understand this, the sooner they will start changing their behavior. Second, let all other team members know that they should include this person from now on. This will help lay down the law that all meetings must be inclusive and no one on the team is to be isolated.

The Social Butterfly

The Social Butterfly on the project is the personality sometimes referred to as the "gabby" or "gossipy" one. They like to gossip and use any bit of information they get. The more you say, "Don't tell anyone," the more you can be sure the information will get out sooner than expected. They're typically perceived as not being as technically skilled as other team members. They're normally well-liked, and you can usually find them talking to someone on the team, no matter what time of day. This leads you to wonder "How do they have any time for work if they're always talking?"

How do you manage this personality? Well, they can be extremely beneficial if you're looking to get feedback on an idea but aren't prepared to publically announce the concept. Getting it into the hands of the Social Butterfly can allow you to get the information out and get informal feedback long before you may be ready to work for that feedback. In addition, befriending the Social Butterfly allows you to get the real "dirt" on the project and what people are saying but may be too afraid to bring to your attention. When trying to plan social bonding activities, this person can be extremely helpful; they can tell you what things people would like and what activities may not work because of one reason or another. Use this person's strengths to your advantage. Integrating this person is usually not a problem as it relates to the social aspect of the project. The real trick is getting them to work on the deliverables they've been assigned.

To focus the Social Butterfly on work, the micro-management technique is really the only thing I've seen that works. They're motivated by praise. This motivation can turn their focus into work effort and deliverable completion. Some may think that pairing the Social Butterfly with our next personality type, the Introvert, is a way to balance out the situation. In my opinion, this strategy only works at parties and company functions. Forcing this pairing to occur regarding deliverables and work assignments can actually have a negative effect where both resources become frustrated with the environment.

The Introvert

The Introvert is the very quiet, extremely intelligent, and analytical resource assigned to the project. They're usually very meticulous and have a strong attention to detail. Not always, but many times this person could fall into the trap of analysis-paralysis and fail to meet deadlines because *all* the analysis isn't complete. As mentioned above, pairing them with the Social Butterfly can cause them to be constantly distracted; as a result, they can take even longer to complete the work. My recommendation is to structure their work assignments in smaller chunks and iterations. I usually schedule their work assignments to one-week durations — very clear and concise direction with frequent checkpoints is essential. You want to give them the time they need to focus and do the work but you don't want too much time to pass without a check-in, because if you let them, they will go off into a black hole of analysis and never come out. Integrating them into the team dynamic is difficult, but can be accomplished by associating their work to work products of other team members. This dynamic forces the Introvert to communicate. It also puts a bit more pressure on them to complete the work in a timely fashion, since someone else is depending on them to finish it.

Steering Committees

As you expand your focus outside the core and extended teams, you'll need to address the subject of steering committees (SC). SCs will often include one or more key stakeholders representing functions such as business, finance, engineering, product support, and sales.

The reason these functions are on the SC is because executive leadership has deemed the project to be either strategically or financially significant to the organization. These members are empowered to make key decisions on prioritization, project scope, cost, and schedule. If the project must deviate from its original objectives, authorization to deviate would also come from the SC team. Additional responsibilities of the SC typically include helping resolve issues and removing roadblocks that are beyond the control or influence of the project team members.

The decision to create an SC for a project or release is usually made by either the project sponsor (the investment body funding the project), or the project core team if they feel they need that type of governance and oversight. The SC are usually the same individuals that authorize and approve "gate" or funding decisions associated with the project(s). Therefore, the criteria that they use to make decisions in SC meetings should be the same.

Most project managers view SCs as necessary evils. This perspective usually comes from the project manager's tendency to take a more passive role in their formation, allowing the SC to form organically. The process to organically form an SC can be one of trial and error as the executive team tries to determine the frequency and relevant content of the meetings. Project managers need to be more proactive in helping executives determine the criteria necessary for these meetings to be successful. By being proactive, the project manager and core team have more control in making sure that this executive interaction is properly incorporated into the natural flow of the project and not based on random intervals (e.g., monthly) that have no relevance on project progress. Done correctly, steering committee meetings can actually be used to help "steer" the project and NOT just be project updates.

When utilizing an SC, a project manager must consider the types of information the SC needs to make approval decisions, as well as the frequency of communication needed. Think of when you need these SC meetings based on when critical information will be available and when critical decisions will be required.

Next, I'll share a hypothetical story to better emphasize what can happen when project managers take a passive role toward SCs. For several years, an organization's project managers were responsible for coordinating, driving, and communicating anything that had to do with SCs. New leadership came in and wanted to change the focus of the business from project to product. Meaning, too much focus was put on the projects and their execution and not enough focus on the product lines and their growth.

Fundamentally, their thesis was sound. One of the ways they proposed becoming more product line focused was to take the responsibilities associated with SCs that were traditionally performed by the project managers and transfer them to the product managers. Conceptually, one could argue that this made sense: if you want to get more focused on product lines, then have the people running those product lines communicate the information. However, what this concept didn't consider was that product management wasn't really skilled nor did they have the resources needed to run these meetings. The concept took the product managers away from their market and the innovation required to sustain and grow that market over time.

The SC meetings turned into templates filled out with metrics that may or may not have been applicable to that specific product line. The project managers resented this shift. They initially tried to offer assistance, but their offer was not welcomed by leadership. So the project management team quickly took the position of, "If you don't want us doing this anymore, then good luck! You're on your own." This position was probably not the right position to take. Instead of creating a better case for themselves and being creative about their involvement, the project managers were soon completely left out of the meeting. The result was that since product management did not have the resources or experience to run the SC meetings, the product line did not grow the way the company leaders were hoping it would. The product managers became more focused on trying to craft the message for the meetings than attending to their product lines. This additional reduction in focus caused certain product lines two to three years of stagnation because they missed their funding opportunities.

Don't be passive when it comes to steering committees! Get involved and stay involved. Lack of leadership on your end could actually hurt your organization's financial returns for several years. Don't let this happen!

If you're passive in your role as a project manager, then you're not a project manager. Project management is about being proactive and actively managing, NOT sitting back and acting as the victim. If you're the victim in your project, then you may want to stop being a project manager and find another career.

The Team Performance Equation

At the beginning of Section 2, we explained the importance of team integration and social dynamics. Using S.T.O.P. as our guiding principles, we broke this topic down into variables necessary for optimum team performance. Those variables were social bonding, team identification, and team orientation. All three of these categories need to be balanced, if you, as a project manager, hope to achieve optimum team performance.

To better understand how the variables relate to each other and to determine when you need to stop the project based on team performance degradation, I've constructed the *Team Performance equation*. This equation is meant to provide guidance and direction to help you determine the different levels of importance each variable has in relation to each other.

Using the same concept in Section 1 regarding the value equation, you need to assign a value to each variable based on the needs of your project. For example, over time, social bonding can have a negative effect on team performance if the team is allowed to get "too comfortable" with each other and become complacent. This complacency can lead to individuals and team leads making project decisions based on personal biases and friendships instead of professional outcomes. Therefore, social bonding, although important, has to be controlled relative to the importance of the other variables.

The equation is:

Team Performance = [Social bonding + (Team identification x Team orientation)] ÷ 2

Team identification and team orientation are multiplied because they have a dramatic impact on each other. By assigning a number to each variable (use 1.0 as the ideal state), you essentially quantify the level of team performance and efficiency necessary for your project. I recommend that you work with your sponsor to assign value to each variable. You'll need their support and assistance with integrating the team to achieve maximum performance. By adjusting the value of each variable as you proceed, you assess your ability and the team's capability to integrate, function, and deliver.

Not all teams are created equal, and you need to communicate with your sponsor immediately when your team's performance starts to decline. The sooner you understand this equation, the sooner you'll be able to structure your plan and optimize project performance!

S.T.O.P. SURVIVAL TIPS

- **Stop!** Project management is not a cookie cutter manufacturing approach; it requires forethought and localization.
- **Think** and understand the type of project you're managing to better orient how these divergent views are going to interplay with each other. Can your project be characterized as new product development, incremental enhancements, or routine maintenance?
- **Observe** the differences in backgrounds, perspectives, and functional priorities that exist amongst your team. Managing team construction and dynamics are a part of your job. They're counting on you to guide them through the project by reducing the risk of interpersonal conflict. *Social* bonding, *Team* identification, *Orienting* around the work, and formal *Planning* are the ingredients to integration and performance.
- **Plan** to capitalize on team dynamics and integration by understanding the *three C's* of your product development lifecycle (conceive, construct, and commercialize) and knowing when to increase communication based on time and team function.
- **Plan** to include in your project schedule business artifacts such as market research, competitive analysis, customer interviews, business case creation, and technology

assessments. This is where you remove the fuzziness from the front end of the project and connect to the business goal of the sponsor/organization.

- **Stop!** If you're going to be a successful project manager, you have to fully understand how the project fits into the overall program or business objectives.
- **Stop!** Remember, to effectively lead a cross-functional team in the delivery of a product, it's vital that project managers have a crystal clear understanding of the purpose, scope, and success criteria for a project. The team is relying on you to be able to clearly articulate the boundaries of scope, the budget, timelines for delivery, and process for delivery throughout the product development process!

SECTION THREE

Communication

Project Planning/Project Designing

Stop! Observe your situation and surrounding environment. Can you proceed safely? Based on the success criteria of the project are your project team members oriented around the work correctly? Have you thought of the best way to plan the project's execution? Think about the number of different industry development methodologies and approaches for solution delivery and apply one (or a combination of several) based on the cultural makeup of your team. Usually, there's not a one-size-fits-all relationship between development methodologies and team culture/maturity. For example, a hybrid approach of waterfall, XP, and Lean development processes may make the most sense for *your* team.

Reach into your tool box and choose the right tools for the job at hand. Any time you introduce a new process to support delivery you need to consider how it's going to be rolled out and how the project will be executed within the development process. Put a plan together that is relevant to you, your team, and your sponsor, in a language everyone can understand.

Keeping consistent with our definition of project management as "the act of managing change against clearly defined project success criteria," we can define project plan creation as constructing the deliverable for how you're going to execute, manage, and communicate change. Creating a well thought-out project plan is where project managers make their money. It's absolutely critical to get it right. In my opinion, project planning is fundamentally where the true value of a project manager can be realized by the team. Project management isn't purely putting tasks, estimates, resource names, and dependencies into a project schedule. To get the most from your team and to be sure your plan is robust and consistent with the goals of the project, you need to do more.

> Using the S.T.O.P. principles, the project plan is the place to capture your observations and data to ensure you've put the right plan in place. The project plan allows you to give context and substance to the project, enriching your schedule with meaning and thought. A schedule without a plan is really just a series of tasks waiting to slip.

I once had a boss who introduced me to a group of coworkers by saying my job as a project manager was to "make sure the trains were kept on the tracks." I had to politely correct my boss and remind him that project management is really about making sure that the track has been laid correctly before the train even gets there. Just making sure the train stays on the track isn't enough, especially if the train is headed in the wrong direction or off a cliff.

Using the S.T.O.P. principles and building on the information covered in Sections 1 and 2, the first thing you must realize is that you need an integrated project plan (survival plan) to be successful. A project schedule isn't a project plan. A project schedule is a tool used to help manage a plan. An integrated project plan should be considered the design document for how the project will be managed and executed.

The bare essentials of any project plan should include the following:

- Quantifiable success criteria for the project
- Project tradeoff matrix (scope vs. schedule vs. cost)
- Work break down (WBS) structure
- Roles and responsibilities broken down by deliverable — RACI matrix
- Communication plan/stakeholder management plan
- Risk management plan
- Change management plan

We've already talked about the concepts of quantifiable success criteria and the project tradeoff matrix in previous sections. However, I want to reinforce, just for a moment, the importance of success criteria in your project plan. Properly constructed success criteria designed around market/business/product needs and project outcomes turn your project plan from one that is inwardly focused to a plan designed around meeting customer needs and ensuring the work done throughout the project has the outcomes expected by your external customers and organization. It's critical that your project plan be built around these criteria and objectives. If it's not, *Stop!* Don't risk sending your team into a mission they can't win — spending significant time and money only to fall short of expectations and overall value to the customer and to your business. Ensuring that your project plan has been constructed based on the project's success criteria is one of the first steps in turning your project plan into a *survival plan.*

Stop Checking the Box!

Have project managers forgotten all their training? Have the basics of project planning all been thrown out the window? Are we as project managers becoming "lazy" when it comes to planning our projects? To some degree, "YES," I believe we have. By lazy, I mean that many current project managers no longer take the time to truly understand the entire context of the work. I don't believe this behavior is by any means malicious or that project managers are intentionally lazy. I believe this behavior is caused by over-allocated project managers constantly trying to play catch up and looking for ways to cut corners just to meet a project deadline.

Typically what happens is a project manager tries to construct a schedule without first creating the WBS. Then, they try to use their schedule to assign resources to deliverables, but fail to think of the other inputs and communication channels associated with each work item and task. In addition, they lose sight of the quality aspect of the project and the capability required to review and approve each deliverable. Finally, they

> As project managers, we sometimes let the increasing organizational and internal pressures impact how much time we allocate to planning. Many times we fall prey to the factors we're trying to control and avoid. If we don't stop and successfully plan our projects, the probability that the project will fail increases dramatically. Most project sponsors want to get to execution as quickly as possible because that's where they get the feeling that "Hey, we're making progress." The entire project team can get caught up in this trap. Unfortunately, in this situation, nobody realizes the true value of planning the project until they experience issues later in project execution.

take the step of defining dependencies between tasks, but because they've skipped the previous two steps, their schedule is, in essence, incomplete. Tasks will get missed, review steps will never happen, and the incomplete schedule will soon start to slip. Even if they do take the time to construct a WBS, it typically occurs at the deliverable/template level with no understanding of what information is really required to complete the deliverable.

Just because a deliverable template exists doesn't mean it's needed or that the entire document needs to be completed. Stop acting in this "check the box" manner. You're missing valuable context needed to help plan your project! Doing what you did on your previous project is not good enough. To do true project planning, especially considering the complexities that go into technology product development, you need information and context to turn yourself from the everyday project manager into the "project tailor."

Project tailoring is when project managers, together with their team, create their WBS based on actual activities and inputs/outputs of information flow, rather than just a predefined deliverable. These tailors then assign the work to individuals, apply estimates and dependencies to the tasks, and construct their schedules ensuring that the resources are fully utilized based on the priority of the work dictated by stakeholder and market expectations. The project tailor typically has experience working through several different types of software development methodologies like waterfall, Agile, Lean, etc., and can pick the aspects from each approach that work best for the particular project. While constructing their schedule, these tailors also create their communication and risk management plans to ensure that all deliverables are in sync and that the entire team is mentally prepared for project execution.

When we talk about the WBS, as project managers, we usually go to some defined set of deliverables/templates that another team or the project management office has created, and we start assigning team members to each deliverable. But do we ever take the time to dive into each deliverable to understand what bare minimum of information is necessary to complete the work and the project? I know this may sound a bit risky and make you uncomfortable. Obviously, the person who created the template was a genius and has streamlined the template so it already captures only the information necessary for your specific project.

Yeah, right. You can see where this is going. Templates are good guides that can help us think critically and ensure we've asked all the right questions as we obtain information. But they aren't intended to be robust solutions capable of capturing the nuances for each and every project. Taking them at face value and requiring your team to fill them out

> One method of tailoring a project is to turn your WBS into a pseudo PERT analysis. Tie your deliverables together through an input/output diagram so your team can see the dependencies they each have. This method will allow a quality check and force a stop decision on which deliverables are required versus which aren't relative to the type of project you're executing. It will also give the team a visual of how dependent they are on each other for delivery and achieving success. An example of turning your WBS into a PERT analysis is presented in the Appendix as Figure A.3.

— so you can check a box that the item was completed — is downright irresponsible. You have to understand that project management isn't about checking a box. It's about context, understanding, and constructing a plan in a meaningful way that will allow the team to execute in the most efficient and effective manner.

Now, I know the philosophy behind Agile is very similar to this and stresses only documenting what is necessary. However, this concept sometimes gets misinterpreted, and teams try to justify that they don't need to create *any* documentation. For most projects, especially large, complex product development initiatives, this path is a recipe for disaster. The question of "to document or not to document" isn't the question you want to ask your team. The question you really need to ask is, "What specific information is required, and by whom, so that they and the project can continue to make progress?" This information isn't only critical for constructing your schedule in the most efficient way, it should also help structure your communication plan. Let me provide an example to better paint this picture.

I was involved with a team constructing a brand new product scheduled to be launched into the market place in 12 months. The team laid out their plan appropriately, conducted the necessary market input and research, and then applied their learnings as they built their market requirements. They wanted to present the customers with a prototype to validate their interpretation of the research. So not only was their schedule aggressive, but their communication plan had to be robust enough to account for internal core team interaction, extended team coordination, executive status reporting (planned and unplanned), and external customer updates and feedback. Keep in mind, this was a new product so the amount of interaction with sales and support staff was also great considering the amount of context and ramp-up time necessary for these functions to support the new offering.

> Lately I feel "done" has turned into a bad word. Project managers seem to be asking if "it's done" without understanding what "it" is or what "done" really means. Project managers need to stop before they check the box and mark a task 100% complete in their project schedules. The indication that a deliverable is complete should be questioned and challenged to ensure the target audience has everything it needs and can move forward without further iterations. The team needs to understand this point and realize the impact of their decision before they call something "done."

As the team iterated with customers, the prototype changed and the requirements document quickly grew — to the tune of over 150 pages — not because there were so many requirements, but because the template was constructed to be everything for everyone. It became the business requirements, the functional requirements, the data flow requirements, the operational and non-functional requirements; it became a catch-all for anything that was captured as feedback by the customer. It became the infamous "req-spec."

The team acted in such a way that the document couldn't be reviewed until it was complete. Once complete, it would be under change control. The project manager then asked the internal team to review this behemoth and set unrealistic timelines for them to provide feedback due to the time pressures of the project. Basically, if they wanted the product out by a certain date, the deliverables had to be completed and reviewed by a certain date as well. This quickly led to a document that was out of sync with the prototype; the document was too large and couldn't be maintained as quickly as the prototype was revised. This also led to an insufficient review of all the material in the document because it was just too complicated and large to review. In the end, release one of the product was incomplete from a market satisfaction standpoint. To this day, questions arise around the original requirements and what was really driving the solution. Did the team really understand what the customers wanted and what would have happened if they took more time and a more incremental review process as requirements were gathered?

What should have been done to avoid this problem? Well, first, the document template should've been reviewed by the team to assess who the internal customers were for each part/section of information required. The customers of each section in the document should have then been asked *what minimum information they needed to do their job.* Based on that assessment, the team could have determined whether it was truly one deliverable or several deliverables that were needed to document "requirements." The project manager could have then taken that information and turned a one-line "requirements complete" task in the schedule into a step-by-step progression of activities, which followed the natural flow of information. That way, each part of the document could be created and reviewed by the appropriate team members as it was completed.

You see, it's not just the deliverable, but the information in the deliverable, that truly matters. Now, I know this may sound like basic information to any seasoned project manager – and it is. But, in today's over-allocated, over-committed world, this is not what's happening.

If we, as project managers, ever hope to improve speed to market, team effectiveness, and overall project outcomes, we've got to start taking a deeper dive into the information required by team members. The project manager got caught up with the pressures of senior executives trying to meet a deadline. She should have said "*Stop!*" Lessons learned go a long way... but only if you actually learn from them. I must confess that I was the one who created that requirements template. Another lesson learned: always let the functional owners of the deliverable create their own templates. They have the necessary context to know how best to construct it. The project management office typically doesn't.

> In general, I believe speed to market is a trap. Although getting products to market quickly is a nice goal, I don't believe it alone wins the game. Many products come to market before the market is ready to accept them or before they have the right combination of features and functions that will appeal to the early adopter types, let alone the mass market. Speed alone does not compensate for value.

Now, as you go through the exercise of determining what information and deliverables are critical to your project, you need to understand who on the team is best skilled to author them, and who is best to review. To truly do that full robust *planning*, you need to establish a responsibility assignment matrix, commonly called a RACI (Responsible, Accountable, Consulted, and Informed) matrix before you start to build your project schedules. The RACI matrix is describes the roles and responsibilities of various teams or people involved in developing and delivering a product. It should also include your stakeholders and any extended team members needed to deliver your project. It's used to help plan and break down the work even further, so that the right people are doing the right work at the right time.

The RACI matrix splits tasks into four participatory responsibility types, which are then assigned to different roles in the project. These responsibility types make up the acronym RACI:

- **R**esponsible – Those who do work to achieve the task. There can be multiple resources responsible.
- **A**ccountable – The resource ultimately accountable for completing the task. There must be exactly one "**A**" specified for each task. The Accountable role is accountable for the creation of the deliverable and the communication associated with the deliverable. The person accountable may or may not also be responsible for the task, but typically it is a person in a functional or team leadership role.
- **C**onsulted – Those whose opinions and feedback are sought. This role is about two-way communication. These are usually individuals who provide approval

on a deliverable. Consulted parties often turn into Accountable parties in the next stage of the project.

- **I**nformed — Those who are kept up-to-date on progress. This role involves one-way communication.

Understanding the different roles and their interaction with each of the project deliverables is extremely important to project clarity — and the RACI matrix is key to helping you reach that clarity. A sample RACI matrix is shown in Figure 3.1.

The matrix is typically created with a vertical axis (left column) of tasks or deliverables, and a horizontal axis (top row) of roles. It's important to note that before a RACI is finalized, it's always better to name *people* instead of just roles or departments to further clarify *who* truly is to be accountable for *what* work. In addition, the RACI matrix helps as a communication tool with the rest of the team. One, you can use it to confirm with individual team members whether or not the deliverables or assignments are correct. Two, you can use it with your core and extended teams to communicate who is doing what and how the interaction and handoff between team members is going to occur.

> Like the project tradeoff matrix, the RACI matrix can be used as a stop sign for your project. If you don't have the right accountability and responsibility mix within your project — *Stop!* Talk with your sponsor; show them the results, and work with them to get the right team in place. The RACI matrix can also be used to help explain to your executive sponsor and senior leadership just how complex your project is. I'm not saying it's the silver bullet in getting that message across, but it can help.

Figure 3.1 RACI Matrix Table

RACI Overview

R = Responsible: Performs work, can be shared
A = Accountable: Only one can fulfill this role, can perform R or delegate
C = Consulted: Those who need to be consulted before final action; two-way communication
I = Informed: Needs to be informed; one-way communication

Role	Product Owner (Product Management)	Business Analyst	Interactive Design	Engineering Lead (Tech Lead)	Project Lead (Project Manager)	QA	Marketing	Sales	Traning	Product Support	Stakeholders	Finance
Deliverable/ Activity	Product Development Stage											
Deliverable 1	A	C	C	C	R	C					I	
Deliverable 2	A/R	C	I	C	I	I	C	C			C	C
Deliverable 3	A	C	C	C	R	C	C	C		I	I	I
Deliverable 4	A/R	C	C	C	I	C	C	C			C	
Deliverable 5	A			I			C	C	I	I	C	
Deliverable 6	A			C			I	I	I	I	C	
Deliverable 7	A			I			C	C	C	I	C	I
Deliverable 8	A	R	C	C	C	C	C	C	C	C	C	C
Deliverable 9	R						R	A			C	
Deliverable 10	R						A	C			C	
Deliverable 11	A			C	C	C		R			C	
Deliverable 12	R						C	A		I	I	

Once you've figured out who's going to do what and who it needs to be shared with, the next logical question is "when?" When does the information need to be obtained, created, and shared? You might think you're ready to begin creating your project schedule. Nope — not yet! What you *are* ready to do is start assessing and designing your plan around optimum efficiency and cross-functional integration relative to each stage of the project and how you want to deliver the solution.

To do this assessment and planning you need to determine what type of development process will work best for delivering the solution and what mechanisms you'll need to control risk. One popular way to manage risk is to introduce project review points. These review points shouldn't only be used *within* your organization; consider also using them externally. You have to think about continuous customer involvement throughout the project to ensure that both internal and external expectations are in line and met throughout the entire project lifecycle.

Review Points Process

The most common type of project review process is the Stage-Gate® process. To arrive at project go/no-go decisions at a stage gate or review point, a review committee will review project deliverables and achievements during the relevant project lifecycle stage and collectively determine whether to proceed to the next stage of the project or to stop and re-factor where necessary. You should use considerations such as strategic fit, technical feasibility, risk, market attractiveness, and competitive positioning consistently when making decisions for each project. The decision criteria used in review points represent the standards for management judgment and decision making and guide the steering committee and senior management when they are assessing deliverable content and quality.

The outcome of each review is a recommendation by the project team to the executive sponsor/steering committee to:

1. continue to the next stage (*go*),
2. cancel the project (*no-go*),
3. redirect the project based on strategic changes or gaps discussed during the review (*rework)* or
4. defer a decision (*hold*).

These reviews are not the same as technical, architectural, or deliverable reviews and shouldn't replace them, as technical and other reviews are conducted before the actual stage reviews. In addition, anything you present during the review should be reviewed first by the core team to be sure each member is in agreement before presenting to senior leadership.

> Each review point gives you a chance to stop and assess the project for viability based on assumptions made on financial models and customer acceptance. If you ever wanted a way, organizationally, to stop a project, the review point process is it. Bring your sponsor into the process; they should be the ones holding the stop sign at each review point until all parties are in agreement that the path is clear ahead.

These are examples of what must be examined and questioned by the committee and project team at each review point:

- Strategic alignment — the project and product fit the business strategy.
- Market of a minimum size exists and will still exist at launch time.
- Technical feasibility is reasonably likely — *risk areas have been identified and solutions are available that can be reasonably acquired by the project.*
- There is a product value proposition.
- You have a marketing plan to prevent building products or product capabilities that lack a "go to market" strategy and plan.
- An acceptable return on risk-weighted investment is available.
- There is an understanding of how the new product offers unique benefits, meets customer needs better than competitive products, or provides superior value. *These estimates should arise from actual tests with customers. How were customers involved with evaluating the product concept or prototype? What was their reaction? What criteria were used to select the best concept?*
- Resources with related experience exist to construct a team that has the scale, complexity, and technologies needed to support the development effort.

- Testing has been comprehensive and robust, ensuring a quality product.
- No "show-stopper" issues exist. *If they do exist, they can be mitigated and there is a mitigation plan.*
- All project and product risks have been communicated and have appropriate mitigation and contingency plans.
- All parties have been informed of and agree with the execution plan and approach.

Based on our high-level product development lifecycle of conceive, construct, and commercialize, you should use the guidelines in Figure 3.2 to prepare and set expectations for each review point along the product development process.

> *Stop!* Review points aren't intended to be informal discussions. They're critical business decision-making meetings used to optimize portfolio performance, scrutinize product development investments, and increase the quality of development execution. You should collect materials for the review and send them to the review committee members ahead of time. During the meeting, you should lead the discussion and allow the core team members to present and support materials in their areas of expertise.

Figure 3.2 Review Points in the Product Development Lifecycle

REVIEW POINT	REQUIRED BEFORE REVIEWS	REVIEW GOAL/EXPECTATIONS
CONCEIVE TO CONSTRUCT	▪ Core team reviews and agrees on gate review documents and sends to review committees for review. ▪ Review committee approves that all contingencies from previous decision /review points have been resolved. ▪ Appropriate legal documents with third parties are in place and compliant. ▪ Review committee members have access to full/appropriate project documentation.	Agree on the understanding of the market requirements and the associated business case and demonstrate a sound project plan and success criteria to support the next stage of the project. Target dates for key next deliverables should be given along with next review point target dates.
CONSTRUCT TO COMMERCIALIZE		Demonstrate that the requested functionality maps back to the original requirements, is technically feasible, and can be tested and achieved in the agreed-upon project plan timelines. Set a commitment date for product launch.
COMMERCIALIZE (LAUNCH OF PRODUCT)		Demonstrate that the product has been built to requirements; present quality recommendations; and ensure that manufacturing, operations, and support of the functionality is understood and accepted.

When considering review points, *stop and think* about what's needed to make those review points valuable. The fundamental evaluation criteria for review points can be grouped into five dimensions: strategic fit, technical feasibility, customer acceptance, market opportunity, and financial performance. These criteria are expected to flex and be tailored to your project type and project phase. For example, you may find you need more evaluation criteria early in the project because the product being introduced is new and innovative, thus implying more risk to its overall success. However, a word of caution... many times I've found that teams and steering committees are likely to question more and be more disciplined concerning evaluation criteria early in the project — when it's viewed as more acceptable to stop a project. Later in the project, they become almost complacent and aren't as disciplined about evaluating the project. My perspective is that steering and project review committees are actually more afraid of what they might find later in the process, when more money has been invested and further expectations have been set.

Sure, it's important to stop projects as early as possible when less time and money have been invested. However, that same level of evaluation, if not more, must occur throughout the later stages of the project as well. Completing a project and releasing a product or product functionality that isn't fully ready, marketable, or customer accepted will prove to be a worse decision because overall bottom line results will be impacted. As a project manager, you can't let that happen. If you do, you'll waste precious time, money, and resources that your organization doesn't have. Allowing a project that should have been stopped to continue can lead to missed opportunities in other areas of the organization, demoralize employees, and possibly damage the company's reputation or brand if the product fails in the marketplace.

On the other hand, you also have to be careful not to allow projects to be stopped just because the organization is afraid to take risks. A certain part of the project portfolio needs to include new, disruptive technologies and innovations to balance the organization. Ensure that projects are stopped for the right reasons based on the evaluation criteria and NOT the weak appetite of your project review team.

These review point guidelines can cause a lot of consternation among team members, so let me spend a bit more time describing how one really should operate within a review point environment. The review point process isn't as straightforward as it may seem, especially when it comes to team dynamics, communications, and outcomes. What I mean is that the review point process is likely to cast a negative dynamic across your project. For example, in the first review point, requirements for the solution typically aren't yet fully understood. This lack of understanding can lead to a team's gross underestimate of the work they must complete. Any unjustified estimates could result in a project that is underinvested. Low investment in the project in the beginning can cause delays in acquiring resources, which can lead to delays in ramp-up time and ultimately late deliverables. This can also result in a complicated and complex message being delivered to a review committee that, especially during the first review point, can quickly dismantle your project. Your team, highly energized in the beginning of your project, could turn into a team facing massive schedule pressures and missed expectations. As the second review point approaches, the team can be perceived as scrambling to "catch up." Stakeholders view this as a lack of confidence within the team to execute and deliver, ultimately leading to project cancellation.

My recommendation for dealing with this situation is to slightly relax the review process – specifically the timing of the first review point. Using terminology like "review point" instead of "stage" is actually very important because "review point" implies more flexibility and adaptability of when the meeting should occur, instead of arbitrarily trying to force fit it into a predefined "stage" in the process. As a project manager, it's important to understand these nuances of communication and control when this first meeting happens. Try to build in enough resources into the front portion of your project to get through a basic understanding of your customer requirements, a conceptual technical feasibility assessment, and any prototyping that can be done.

I know this isn't easy, because what I'm essentially recommending is that you ask your stakeholders to make more of an investment up front and "go" with the project, at a time when they typically don't feel comfortable and are more inclined to *stop* it. You have to convince them that lack of investment in the beginning may cause this "great new project of theirs" to be killed unexpectedly. You want to avoid that risk. Remind them that you're on their side and you don't want a review committee to question the viability of the project. You want them to be enthused about how great this initiative is going to be, and the only way you can do that is to invest slightly more than usual upfront. Trust me, this approach of greater investment in the beginning of the project will get you the data you truly need to validate whether the project

> Remember, as a project manager, your job is to ensure that the project is viable, and if it's not viable, make the recommendation to stop it. The only way you can do that is to have enough data to support your recommendation. It's better for you to make that recommendation than for senior leadership to make it for you. Stopping a project is only bad when it's stopped for the wrong reasons.

is viable AND allow your project to be killed *for the right reasons* – not the wrong ones.

And in this case, the wrong reasons are that due to lack of resources (time, staff, money) the team wasn't able to get all the information they needed before meeting with the review team. Give your team that time and make sure your schedule supports your argument to ask for more time and money.

It's important to note that once review points are established and communicated, review point dates are immovable. Meaning, even if all deliverables for that review aren't complete, the review still needs to occur. Based on project status, it'll be up to the review committee to assess whether the team needs an additional review meeting OR if the team can proceed with conditions that must be met by a certain date. This immovable date concept may sound a bit constraining from a process standpoint. But remember, the review points process isn't meant to replace or substitute for an Agile- or Lean-type development process. Review points, by definition, are meant to act as business decisions, which have a time component. Every project has a point when the costs outweigh the benefits. You have to allow your stakeholders the ability to make those decisions. The following story is an example of what could happen when the project manager doesn't take control of the review points process and allows the organization to predetermine it for them.

During the early stages of a previous employer's organizational development, I was part of a team that was crafting a product development lifecycle. This lifecycle was meant to create a common lexicon across the organization for communication and managing expectations on the progress of project delivery. This lifecycle took the shape of several stages with defined deliverables. Information technology leadership made a rule that true commitment to delivery dates wouldn't be set until we had proven the project was technically feasible and had completed analysis deliverables — essentially all work before code was written.

In theory, this positioning made sense, and most of the organization supported it. However, because we took this hard-line approach to stages and an even less flexible approach to delivery commitment at each stage, project delivery teams quickly became confused and began struggling with delivery deadlines. Not all projects fell neatly into these predefined stages. Some projects had to be broken into "chunks" or iterations. As a result, some parts of the project were ready to move from analysis to development while other parts were still being analyzed. This situation created a lot of confusion, especially when trying to communicate the status of the project to leadership. Project teams tried to adapt to the new protocol but either went too far to the left — stopped doing gate reviews completely — or too far to the right — stopped communicating and committing to any deliverable or any date until ALL the analysis work was complete. Ultimately, leadership re-established the structure and we found a happy medium, but up until that point, our execution and delivery metrics were horrible because we had no insight and were waiting too late to communicate important information regarding the projects.

Agile vs. Review Points

Over the past several years, a number of development methodologies have received a lot of notoriety. This attention has caused conflicts among team members as they try to determine the differences and benefits of each methodology. I've seen this conflict quite clearly in larger organizations where a review points process has been instituted and management has required that all projects proceed through it. The conflict occurs when the review points schedule is predetermined and out of sync with the rest of the project. The problem is exacerbated when a software development team uses a variation of Agile development and views their work as not falling so neatly into a step-by-step review/waterfall approach.

A development team can quickly become frustrated and confused when asked to prepare for a review point that they feel they're not ready for or that they view as not being Agile. Their feeling of unpreparedness isn't caused by being behind schedule, but rather by their lack of readiness to communicate what they think senior leadership wants to hear. If the review point is allowed to proceed with no course correction or communication intervention, the team and senior leadership can find themselves in further conflict because the team believes that the review points process is constraining and that leadership doesn't understand the natural process of software product development. The team feels set up to fail with no way of ever meeting management's expectations. This result is especially common when these organizations further confuse terminology between the different processes and try to force fit the review points process into a development methodology.

Planning a project with these two perceived process conflicts can be tricky, both from a pure planning perspective and a team dynamics perspective. My recommendation is to first *stop* and *think* about what type of software development process your team is *actually* using. They might be calling something "Agile" when it fact, it isn't Agile at all. Agile seems to be turning into a generic term; some project teams are trying to use Agile before they understand its principles and how to use it.

Let me further explain this point with an example. I had just recently started a new project and was introduced to a team that apparently was using Agile. I had arranged a meeting with the engineering director to better understand the inner workings of the project and how I could help. The director was quite proud of the team's progress and touted their use of XP. I had

> *Stop!* The review points process is a business process used to control risk and inform decision making; it is NOT a development process. Development processes like Agile, Rational Unified Process (RUP), Lean, etc. are tools or methods used to design and develop products. You must determine when review points should occur using a bottom-up approach based on when information is available. This will allow stakeholders and senior management to make business decisions on the project. It's not about a review points process versus a software development lifecycle (SDLC) — in this case a variant of Agile development. It's about how the review points process and a SDLC must work together to meet the communication needs of senior management and the development flexibility of engineers and product development.

a lot of experience with XP so I began to probe on their use of it. I started with some basic fundamental principles, thinking I would build additional questions off of their responses. I first asked about how they were using the immovable milestone concept and how the team was adapting to those dynamics. The director responded by saying, "Well, we really aren't that rigid in keeping the milestones fixed." Hmmm, I thought... So I tried again with the principle of paired programming by saying, "I see the project has been going on for two years now and the budget is quite large. How did you convince senior management to let you double up on resources to do paired programming?" Again, the response was vague: "We really couldn't afford to do paired programming."

By this time, I'm sure the director noticed I was looking puzzled as I couldn't find any evidence of XP. With one last attempt, I tried to give him the benefit of the doubt by asking a question about project tailoring. I thought maybe, just maybe, they had taken some of the principles of XP and tailored them for their needs. I said, "I understand you can't afford twice as many resources to support paired programming, but have you tried having your QA resources sit with your developers as they begin their designs so that they can at least question the design based on their understanding of the requirements and get a jump on the QA test case construction?" Their response was, "That's a good idea. No, we never thought of that."

The point of my story is what they were calling Agile and XP was more based on the burn-down concept and trying to plan work using story points. But they were not using full-blown XP. And because they simply threw out some basic principles without trying to figure out how best to use them, their project was quickly unraveling. In fact, it wasn't too much longer after that conversation that the team had to present the CEO with a new base-lined schedule and defend their need to double their project timeline from two to four years. Not a good position to be in.

As a project manager, you can't let a team get this far in the process before trying to correct their course. If the team in the above example had used the S.T.O.P. principles, the call to the CEO's office could have been avoided. Plan your project appropriately and don't cut corners on process. If a team wants to explore Agile development, then do it as a team and verify that the team interaction and delivery requirements justify executing the project that way.

Don't let your project get stopped because you didn't plan it correctly. Use the S.T.O.P. principles to guide your team to a safe and effective delivery methodology that is backed up with facts and hard data, not opinion or passion. Otherwise, your project will end up like the above example: stopped because senior management lost confidence in the team's ability to plan and execute. Once you lose senior management's trust, it can be nearly impossible to regain it.

Choosing the Right Process

I was recently talking with a product manager about general product development practices. This person had convinced others that they were well educated in the ways of Agile development. As we were discussing the finer points of life and how we would change things if we were rulers of the world, he brought up the topic of scrum — a type of Agile development method. They referred to scrum as a project management framework, not a software engineering framework. I found this interesting, because I've always considered scrum a software engineering process.

I mentioned that in the world of product development where your cross-functional team includes resources from sales, marketing, training, etc., the entire project can't be solely run through scrum. The software development portion can, but the entire project can't. It just isn't feasible to construct the fuzzy front end of product development, as well as the product launch and commercialization activities, in a scrum fashion.

Needless to say, our "discussion" turned into a bit more of a debate, as we wrestled with the details of individual deliverable creations and cross-team communication and integration. My story shows that different people have different perspectives and call things by different names. Fundamentally, I think people and teams get too hung up on the latest cool thing and get lost in the name and its applicability. As a project manager, knowing how someone is defining something is important for context. However, it shouldn't make a difference when constructing a project plan and schedule — IF you construct it based on what information is needed and when that necessary information will be available for the right team members to progress and launch a product.

End users and customers don't really care if you ran the project using Agile — or any other particular methodology. What they do care about is the usability, quality, and timeliness of the resulting product. This is how you need to think when constructing your project plan and schedule to be the most efficient and effective way to deliver the end product. However, the better understanding you have of different processes like Agile, Lean, etc., the more tools you'll have to help you meet those requirements.

As you begin to think about which process makes the most sense for your team to execute and deliver against, be realistic — identify your risks and choose a process that mitigates these risks and helps you deliver a successful product. If your project's risks (both internal and external) don't match the process, stop trying to make it fit, and choose another one. Your customers and their expectations are an important part of your risks. What are their needs? What is their experience? How much are they willing to work with you? Choose a process that will minimize those risks — whether they are date to release, defects, features, or cost.

Determining the true drivers and constraints of the project, identifying what constitutes "success," and understanding the current risk profile, will all help you determine what process is right for you. You may find that none completely addresses your organizational needs out of the box and that using a hybrid or blended process is really the answer. This is the point when you begin to revisit the idea of becoming a project "tailor" as we mentioned earlier in this section. To get the right process for your project, you might need to evaluate and combine the characteristics of multiple processes.

Let's take the example of a development team suggesting an Agile-type process. Say you've concluded that scope (features) is the driving constraint of the project and where the least options exist to trade off. You've also concluded that your market requirements are not changing rapidly. So, the scope is pretty cut and dried and it's clear what value needs to be delivered to impact your market.

From a product delivery and launch perspective, Agile would not be the most ideal process because requirements are not rapidly changing. However, a hybrid process may make the most sense. It may, in fact, make sense to use Agile for only the development and QA processes so that other business representatives like product management, sales, and marketing could get quicker feedback back to the team on what was being developed, and then correct the course where appropriate. Processes like two-week iterations and reveals on feature/functions are perfect for this rapid type of feedback, because scope was the most important driver constraining your project.

This example would be the perfect case for using a hybrid waterfall/Agile approach. Here's how it would work: the waterfall portion could be used for market definition and market requirements, and perhaps even for more detailed requirements or functional specifications that someone like a systems engineer or business analyst would create. The process could then become very Agile-like and run through a series of iterations and reveals through the development and verification phases. Finally, the process could turn back to waterfall for end-to-end testing of the entire system and all of the feature sets that are about to be delivered to the customer. All of this tailoring is completely acceptable if it meets the needs and capabilities of the team and accomplishes the project's goals.

The following story is an example of how tailoring could work when your organizational delivery methodology is waterfall and your team is Agile. Earlier I told a story of a rigid review points process and the conflict it caused with project delivery teams. There's even more to learn from this story. During the process, one team took a stand and decided to tailor their project, their governance, and their communication — all for the better. They actually blew up the entire gate review process for their project. Instead of saying they didn't want any stakeholder involvement, they used the steering committee concept and said, "We would rather have monthly steering committee meetings; based on project progress and status, where we would determine collectively what "stage" we were at and what "gate" we were trying to pass.

They broke their project schedule into iterations, or chunks, of functionality and communicated each chunk status accordingly. The process worked great because the team had the flexibility they needed to work more in a natural, Agile-like manner, and stakeholders and leadership better understood the context and components of the project and delivery approach because they were communicated and collaborated with on a more frequent, monthly basis.

In the end, *what matters most is context and understanding*; understanding the types of planning options available to you and ensuring that the tailor approach is thoughtfully selected. This leads to the best outcomes for everyone — your team, your customers, and your organization — and gives you the greatest chance to succeed.

Communication – A Constant Public Relations Campaign

Project communication is often under planned and causes much overreaction. Typically, project managers who fail to create a plan fall into the trap of having the "who," "what," and "when" of project communication dictated to them. They fall into this trap not because of some organizational dictatorship, but quite simply because without a plan, the sponsor or senior executive has a right to take action and cover their vulnerabilities so that at least what they want communicated gets communicated. This leads the project manager down a path ranging from incomplete communication when the project is going well to misinformation when the project is becoming troubled.

This perspective may sound a bit crass, but it's often reality. Most people, especially organizational leaders who have invested company dollars in the project, want to hear good news about the project being on track. Very few ask the hard questions that can lead to "bad" news or information that the project has gone off the rails in some capacity. This "only good news" routine can create a culture of "spin doctors" resulting in more work for you and the team as you start to spend more time crafting the message instead of reporting the facts, managing expectations, and focusing on the real work of delivering. The more communication channels there are and the more complex the project, the more spinning that can occur, and the more communication can get out of control.

When thinking about project communication, it's not only important to think about the number of channels of communication but also the way the communication is delivered to each type of customer. The most successful people in project management seem to have a knack for translating technical information into something very digestible for the masses. And even though many say that 90 percent of project management is communication, I don't think that says enough. To be a good leader or project manager, I think you also need to add science to your toolbox. When I refer to science, I mean we should be objective and data driven — having the ability to 1) determine what items or deliverables are going to serve as feedback mechanisms to allow the team to make changes along the way and 2) how you're going to communicate. Please remember, you must base your communication on data — not your opinions, passions, or what you think today. Any type of communication around a project, product, or portfolio must be a data-driven conversation, if it's to hold any weight. Without data, you'll find yourself in the uncomfortable position of not being able to answer questions, not knowing what next steps to take, and potentially generating a lack of confidence from senior management. Before we discuss how to construct your communication plan, let's review the S.T.O.P. principles, making them relevant to building such a plan.

- **Stop** before you act. Don't rush into any communication until you understand how the message will be received.
- **Think** about what you want to communicate and the goals of the message. What is the right way to safely communicate the current state of the project so that

the message is just in time for proper decision-making to occur? Make sure the messages are aligned with your sponsors' goals and expectations. If you have not defined those key take-away messages, be sure you do so before reporting on the project. It's critical that your stakeholders, sponsors, and team members are all consistent and speaking with one voice regarding your key messages.

- **Observe** the situation and surrounding environment. What are the risks of over- or under-communicating? How can you reduce them? Put together a communication plan that is relevant to you, your team, and your sponsor.
- **Plan.** Develop your communication plan to reduce the risks and decide how to best implement the plan safely and with transparency. Ensure proper feedback mechanisms are put in place so you can monitor how the communication was received and whether you should alter the communication in the future.

Before we get into the elements of what should be included in a communication plan, let me share an example of what can happen when a communication plan isn't created.

Senior leaders at an organization I was involved with constructed an initiative without any project management involvement. The theme of the initiative was convergence... or so some thought. The senior managers had given it a fancy name and created a slide presentation with business case assumptions, functionality expectations, and scope that extended across the entire organization.

The initiative, which turned out to be an entire program of effort with multiple projects that had multiple releases, had gone through the approval process gauntlet. It was then announced, to the entire organization, that it was ready for execution. Next, "leaders" — not including a project/program manager — were assigned to it. Those appointments were then announced, and we were off to the races. Now, you may be saying to yourself, "This doesn't sound too unusual, at least senior leaders had come together to completely endorse and stand behind the program. Sounds like it was communicated to multiple levels within the organization, very early in its lifecycle — so what's the problem?" Well, the problem was that no one really understood what the program was or the value it was intended to deliver. Some thought the value was convergence, others thought it was new capabilities, and still others thought it was a complete technology exercise that was designed to cut costs and product maintenance.

Because a proper communication plan hadn't been put in place and only partial information was communicated, we had a situation where the entire organization had created many interpretations (with just as many expectations) of what the goal of this program was — to the point of confusion. This highly energized initiative turned very quickly into a communications nightmare where the fundamental question being asked by everyone was "what is it?"

So what did we do to address this situation? Well, first we assigned a program manager. The program manager assessed the situation and immediately began to create the communication plan. We had the communication plan approved and spent the next six months, essentially, on an internal public relations campaign. We designed the plan to reset expectations to the senior leaders that had already approved the program. (That was "fun"!) Next, we broke the organization into smaller groups and communicated what the realistic goal of the program was, what it meant to them, and what communication they could expect over the next several months.

Was this all bad with irreversible outcomes? No, not at all. However, because a communication plan was not established from the beginning, the amount of effort needed to re-communicate the program to the entire organization was enormous and took time away from the entire team — time that they could have been using to actually work on the projects.

Remember that proficient project management is about *both* science and art. Science is the ability to create a hypothesis, determine a means of collecting data to either support or refute that hypothesis, and then analyze the results and make a decision based on the data. Apply science to your project and *stop* to ensure enough data exists before starting. There's a lot of passion and emotion when a new project is kicked off. Don't kill the passion, just remove the emotion from the decision-making process and get the necessary data to ensure the project thesis is sound and viable.

Taking the S.T.O.P. principles into account, let's focus on the **S** and **O** principles above and combine them — *Stop* before you act so you can perform an *Observational* analysis of the situation. This will give you the context you need to properly communicate. Perform this observational analysis by taking into account your situation and by understanding your stakeholder's perspectives. This analysis is translated into your communication plan by providing sections within the plan to address the following:

- Situational analysis: Briefly describe the issues, problems, and circumstances that the project is trying to address.
- Stakeholder analysis: Assess the level of stakeholder impact, required awareness, and specific concerns. You can categorize stakeholders into the following groups:
    - Involve extensively
    - Keep informed
    - Enlist as needed

Next in the S.T.O.P. paradigm you need to **T**hink about who needs to be communicated with, any constraints around the communication, and what outcomes you hope to achieve with the messages you deliver. You should build several more sections into your communication plan by addressing the following:

- Target audiences: Identify *who* is the receiver of your communication. Be sure to include both internal and external audiences.
- Communication messaging sources: Identify the key people who will be delivering messages. All communication may not be coming from you. They may come from your supervisor, general manager, sales, etc.
- Communication specifics: Discuss more of the specifics or boundary conditions of the communication. This

includes frequency, level of detail, and limitations on who is allowed to receive the communication. Create a plan that is spread out enough so that you don't overwhelm your customers with communication, but rather provide them just enough to stay informed and stay motivated with the initiative.

- Communication channels: The world of social media has changed so much in the past five years. Think about the different channels for delivering information. Some messages are better face to face, while email is perfectly acceptable for others. There is also the option to host webinars and webcasts. Explore all channels to determine what is the best combination for your customers.

Finally, plan the story you need to tell. Expand your communication plan to include the following:

- Key messages — internal/external: Create the team and management story. Remember, your project is trying to solve a problem, issue, or concern regarding customers — which could be both internal and external to your organization. Everyone from the top down needs to be consistent with what and how they're communicating.
- Marketing approach — internal/external: Be objective and keep the positive news flowing to the right people. This is where the public relations campaign sets in. Internal postings, blogs, marketing campaigns to customers, etc., all play into your communication plan.

- Project reporting schedule: Decide how often you should report project status. Every week may make sense in some cases, but not others, depending on the type of project and how quickly progress can be made.
- Evaluation and measurement: Clarify how you're going to measure and evaluate the success of the project. It may be by traditional means — against scope, schedule, and cost. But a word of advice: focusing your evaluation and measurement around success criteria and the business objectives that the project is trying to accomplish will ensure better alignment with the team and stakeholders. Customer satisfaction and retention, ability to meet service-level agreements, and turnaround time to address production issues are all examples that may be a part of your evaluation and measurement. It's important to communicate what you're going to measure, why you're going to measure it, how often it needs to be measured, and who will be affected. Make sure the data you're collecting will directly impact decision-making.

Figure 3.3 represents a high-level visual of how data flows across your organization throughout a product's lifecycle. A project manager's focus is usually only on the project/ program metrics. However, understanding how information flows across the organization can be extremely beneficial in determining how the project metrics fit in with the rest of the metrics your organization is collecting. This understanding will help you structure your plan to collect data and measure your project's progress.

Figure 3.3 Metrics Process Across the Product Development Lifecycle

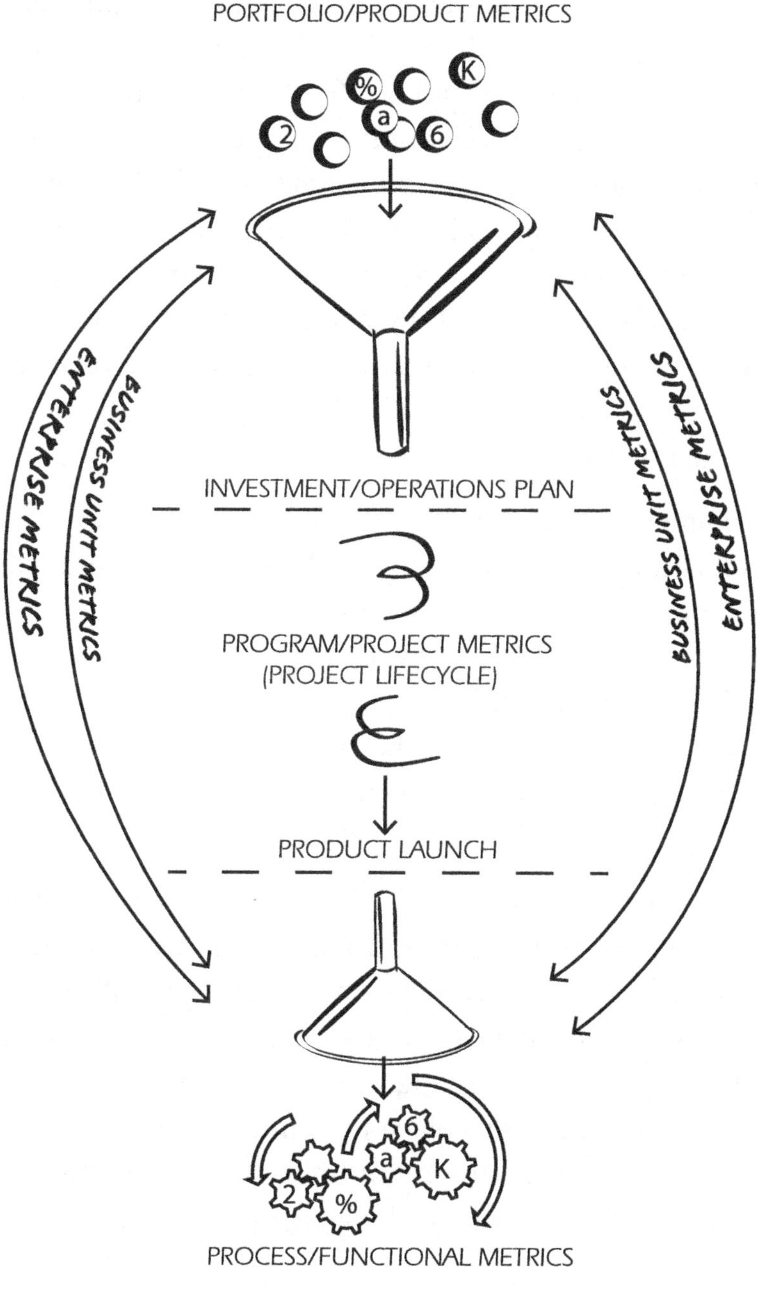

Metrics

Metrics has become somewhat of a bad word in the industry. It seems that we went from not having enough data to being overwhelmed with information without a clear way to decipher it. One bit of advice to all project managers out there: strive to become *the analytical advisor* to your project and organization — not just the *expert advisor*. What I mean is don't expect the organization to trust you because you're really, really smart (the expert advisor) and you've been doing this for years and they should listen to you... NOT! Become the person your team and organization turn to for the facts and analytical data supporting the project — the analytical advisor.

To become that analytical advisor for your project, you need to understand what data you need to help you and others make *informed* decisions during the course of its lifecycle. Determine what relevant data and information will encourage and enforce critical thinking by categorizing metrics based on their value and purpose. Going back to your project success criteria, determine what measurements are needed to verify whether your project is on track or at risk.

Now, I know every project is different and each will have its own requirements for information gathering and data collection. However, I've developed a framework using the S.T.O.P. principles that can be applied to any project/program initiative. This list is intended to ensure you have the right feedback mechanisms in place to eliminate any gaps in your knowledge, understanding, and ability to communicate the true status of the project, in a language your stakeholders and team will understand. Let's try to make the metrics process simple and intuitive by breaking down metrics into the following types:

- **S**topping metrics
- **T**racking metrics
- **O**bservational metrics
- **P**rocess/**P**lanning metrics

If you're contemplating collecting any other metrics or data beyond these, I would advise proceeding with caution. Metrics outside of this list typically fall into the category of *informational;* informational in the sense that you have the ability to collect the data, but it has no relevance to decision-making for your project. Informational metrics can lead to situations where your stakeholders become buried and confused by too much data and can't determine what to do with the information. They will typically look at you with glazed eyes and ask questions like, "So, are you telling me things are okay or that things are bad?" Let's look at these metric types in more detail, starting with *Stopping metrics,* to further explain their applicability.

Stopping metrics are the most important feedback mechanisms used to understand when to "stop the line" or stop the project. The phrase "stop the line" comes more from a manufacturing environment where you need to actually stop the manufacturing line because something has gone wrong or something unexpected has happened. Stopping metrics specifically show the conformance to or deviation from previously established benchmarks or standards that the project needs to execute against for success. Ensuring that all project success criteria have been met and tested before a project is released to the marketplace is a great example of a stopping metric. If all criteria had not been met, you would stop and not proceed to production because key features were missed and your business case was out of sync with your product. Other examples of stopping metrics are:

- If the product can't ship without completing the construction of five features, measuring the number of features is an important metric.
- Going over budget or forecasting over budget by a certain amount from the project's original business case may prove to have an unfavorable ROI or IRR.

Introducing data integrity issues that cause your customers to make incorrect decisions about their organization is project failure. Running your project out of compliance with standards set by independent governing, audit-type bodies is project failure. You want to use these types of metrics to reduce risks to your project by getting constant feedback on parameters along the way. Identifying these metrics first ensures that the right attention and focus is applied to the right areas of your project moving forward.

Tracking metrics are the second most important type of metrics. Tracking metrics are specifically designed to track the current progress of a project. They give you the insight you need to evaluate the health of your project. If you were a doctor, these metrics would be your patient's vital signs.

Examples of tracking metrics might be graphs showing the amount of work completed or tables showing which features are in development or what parts of the project haven't been started. They can be used to show how much budget has been spent to date, or what relative velocity the team is performing at. These kinds of metrics are generally more immediate than other types, and show a snapshot of the project in its current state. They supply the data you need to be proactive and course correct where necessary.

Observational metrics are used for comparison purposes. They allow you to use historical information to sanity check your project, or parts of the project, against previous versions or similar work completed in the past. Having any type of historical information early in the planning process can be extremely beneficial in setting some initial expectations of how the project will execute. But be very careful; every project is different and what seems very similar at the beginning may prove to be quite different as you go through execution. I would say these types of metrics help with general understanding, but the historical context of how they were gathered must be fully understood if they are to be useful to the current project. Although valuable, out of the four metric types, I consider observational metrics to be the least important to your project.

Process/Planning metrics show the expected state of the project, or project components, at a future time or they can show how effectively a current process is operating relative to organizational planned goals. Examples are sales forecasts after the product has released, projected cost reductions or operational efficiencies planned to be gained after project completion, graphs showing the projected defect counts during an ongoing project or customer defects over time for a project that has shipped. These types of metrics are typically used to try to predict the future. Although prediction metrics are typically very "fuzzy" and subjective by nature, they're very useful to compare against the current tracking metrics to determine whether things are going as expected and, if not, how far off they are.

For metrics to be useful, everyone involved with data collection and metric generation must be using the same data and getting the same results. For example, when counting defects across the various components in a project, it's important that everyone involved in the counting has the same understanding of what constitutes a defect as opposed to a design change. If they don't, the numbers will be off and the metric will show incorrect results.

The only way to ensure common understanding is to fully document the definitions and process for deriving the metrics and have complete clarity for the data that feed into the metrics. It's important to document all of this in a common place and ensure that everyone involved reads and understands it before the data collection begins. To help with this process, Figure 3.4 shows an example of how best to describe, define, and collect metrics that may be important and useful to your project/program/portfolio, ensuring all are aligned.

The metric itself is the most important part, but if it's reported too frequently or not frequently enough its usefulness will be lost. Key metrics should be reported often; a failure or variance out of range in one of these areas that's discovered too late in the project will ultimately be very expensive to mitigate.

Figure 3.4 Metrics Communication Table

Metric Category	Metric (Description)	Metric Creator	Leading or Trailing Indicator	Economic Benefit/Value	Threshold/ Tolerance	Can we measure?	How often do we calculate?	Do we capture this now (Y/N)? If so, where?	Level of Effort to Capture	Customer of Metric (Owner)			
										Executives	Senior Mgmt	Project Core Team	Extended Team
Portfolio Management													
Program/Project Management													
Development/ Operations													
Quality Assurance													

The entire point of including metrics in your communication plan is to avoid any surprises to your team and, especially, your stakeholders. Use science and data to become the *analytical advisor* to your project and incorporate the right level of metrics in your communication plan so all team members and stakeholders know when to *stop*.

Stop! With All the Lists

If you ask your typical project manager what area they could be more proficient in, they will usually say risk management. Maybe that's because the process by which we manage risks has become such an overdone science that most teams, especially on the business side, can't relate to what we're trying to do. I can't remember anyone ever saying, "I can't wait to go to my next risk management meeting."

In fact, out of all the project management meetings a project manager sets up, it's probably the risk management meetings that the team is least enthused to attend. Let me try to break down the science of risk management a bit and ease into it before jumping into the calculations used to quantify a risk. For that, let's do two things. Let's first revisit S.T.O.P. and tailor it to risk management. Then, let's look outside the industry to where real risk management is being done and lives are at stake — the military and health organizations.

- **Stop** before you act. Don't rush into a situation without considering the risks versus the benefits.
- **Think** about what you're about to do. What's the right way to safely accomplish the task? What risks need to be mitigated before you start the project?

- **Observe** the situation and surrounding environment. What are the risks? How can you reduce them?
- **Plan.** Develop your plan to reduce the risks and decide how to best implement the plan with safety and transparency.

If you think about it, risk management is a part of everything we do. Risk management is the process of identifying and controlling hazards and making risk decisions to protect yourself and your team. It's applicable to any project or environment. The following is a summary of the Five Steps to Risk Assessment, from the Health and Safety Executive, a British independent watchdog group.[5] This five-step process for risk management (outlined and illustrated in Figure 3.5 on the following pages) is essentially a synonym of the S.T.O.P. principles. If you start your teams off with this approach, you'll add value and make the process of risk identification and management much more relevant. Then you can do all you need as a project manager to further quantify the risks for the team.

Let's start with this five-step process before we move into the quantification process.

1. **Identify hazards.** Consider all aspects of current and future situations, environment, and known historical problem areas.
2. **Assess hazards to determine risk.** Assess the impact of each hazard in terms of potential loss based on probability and severity.

[5] Health and Safety Executive. *Five steps to risk assessment.* Retrieved from www.hse.gov.uk/risk/fivesteps.htm.

3. **Develop controls and make decisions** that eliminate the hazard or reduce risk.
   - Reassess hazards given the controls.
   - Determine the proper decision authority.
4. **Implement controls** that will eliminate the hazard or reduce its risk.
5. **Supervise and evaluate.** This enforces standards and controls. Evaluate the effect of controls and adjust or update as necessary.

Figure 3.5 Five-Step Risk Management Process

Starting your team off with a conversation of the five things you, as a team, need to do regarding risk management is much easier than having them look at some spreadsheet where they're trying to figure what a risk priority number (RPN) is or why they should care. Focus yourself and your teams on these five steps. I have created a graphic, "The Five Steps to Risk Management" (Figure A.2 in the Appendix). You can hang this in your cubicle or office as a reminder to keep the process simple and focused.

Sometimes, as project managers, we find new ways to fall in love with spreadsheets. We wind up creating more work for ourselves and our teams, and we start to live up to the perception that project management is all about "overhead" and process and provides little or no value to the team. We keep separate lists, never incorporating the work into the project schedule, thereby never truly assessing their impact, and quickly losing control of the project. The example that comes to mind is when I find a project manager discussing their issue-tracking spreadsheet with the team. Typically, the project manager keeps this issue-tracking spreadsheet separate from all other tools they're using to manage the project. When this occurs, I usually start a conversation with the project manager by asking, "How are these issues affecting your project?" They usually start to struggle and then say something like, "Well, we are tracking and assessing the issues and *then* making changes to the schedule." *Really*? Shouldn't the assessment be a part of the schedule? And if it is truly an issue, should the tasks requiring you to manage it also be in the schedule as well? Let me now try to dispel this practice of keeping a separate issues list that is independent of your project schedule. This just might save you and your team from looking at yet another spreadsheet.

Let's go back to the fundamental definitions that a well-trained project manager is usually armed with. Issue management addresses the process of identifying and disposing of obstacles that block a team from achieving its goal. Risk management deals with increasing the probability and impact of positive events and decreasing the probability of events adverse to the project.[6]

From my perspective, proper issue management isn't being done correctly in our projects. When you manage issues independently from the rest of the project, your ability to assess the true impact of the issue is degraded. Future events either have the ability to impact the project, thereby needing to be assessed and managed (risk), or they don't, in which case false positives need to be ignored. Therefore, why have two separate spreadsheets — an issue log spreadsheet and a risk log spreadsheet — if they're both capturing the same event. Ask yourself if the "issue" is really a risk, and if so, manage it within your risk management plan. Allow the assessment to occur and incorporate the management of that risk into your schedule. If the issue is currently happening, sometimes referred to as a triggered risk, then the "issue" should be managed within your project schedule as well. Managing a real-time issue outside of the project schedule doesn't really allow you to "manage" it. It only allows you to report on it and ask questions about it in your weekly status meetings. Either get it in the risk management plan or get it in the schedule. Keeping an issue on a separate issues list will set you up for failure.

[6] Project Management Institute. *A Guide to the Project Management Body of Knowledge:PMBOK Guide, 3rd edition*. Newtown Square, PA: 2004.

So often, we as project managers forget to *stop* and understand the conversation before running to our risk register. How many times have you been in a team meeting when near the end of the meeting a team member unloads a barrage of information. It's as if the team member had been silent for so long that they just couldn't take it any longer and they verbally exploded. The meeting runs out of time and what do you do? You run to your risk register and start inserting all the risks that the team member stated, right? We have blank spots to fill in our risk register. We now have information to fill those blanks with, right? Wrong! *Stop!* You don't have enough context yet to determine whether the items brought up are even risks. Set up a meeting with the individual to assess and better understand what their concerns are and determine a plan going forward. If risks are in fact identified, then assess the risks appropriately. If they are just concerns, please don't run and put them on an issues list! Work through the concerns with the individual and get them back on the same page with the rest of the team.

The following paragraphs further describe this risk management process. Let me start with some basic definitions:

- Risk — An event that may occur with uncertain likelihood and timing. If it does occur, it can have an adverse impact on the project.
- Cause — An event or condition that could precipitate the risk and is considered in determining preventive actions.
- Preventive action — A task that is incorporated into the project plan and completed to prevent or reduce the likelihood of the risk's occurrence.

- Trigger — A timely means of detection or notification that a risk has occurred.
- Impact — The effect that must be dealt with if the risk occurs and that is considered in developing a plan of contingency actions.
- Contingency action — Tasks that are executed only if the risk occurs and that reduce the risk's adverse impact on the project or otherwise help the project recover from the risk. These can sometimes be called "work-arounds."

The first step in risk management planning is *identifying potential hazards or risks*. Identification can be achieved by using techniques such as:

- Team brainstorming
- Interviewing subject matter experts or other appropriate individuals
- Considering, in addition to scope, schedule, and costs, the following areas to identify risks:
    - Software complexity
    - Materials and material acquisition
    - Allocation of capital equipment
    - Vendor relationships and contractors
    - Complexity of data
    - Having the right key subject matter experts available and allocated appropriately
    - Familiarity with new technology

- Resources available for aftermarket support
- Sales channel development
- Competitive threats in the market

To really conduct a risk or hazard assessment, you have to remember to use the five-step process and consider the environment — both internal and external to the project team and organization.

Once you have identified the risks, your next step is to *assess each hazard or risk.* What type of hazard or risk is this? What loss will the project experience if this risk is realized and turns into an event? To do proper risk assessment, you need to prioritize the risks based on their likelihood and impact. Risk assessment is applied to each specific risk event individually rather than the project as a whole. Typically, it's largely intuitive and based on the experience of the team and their insights into the nature of the project.

Use the following questions to assess each risk:

Likelihood — What is the likelihood the risk will occur?

- High (5) — It's expected or very likely the risk will occur.
- Medium (3) — It's somewhat likely the risk will occur.
- Low (1) — It's unlikely the risk will occur.

Impact — To what extent will the project scope, schedule, and/or cost objectives be negatively impacted if the risk occurs?

- High (5) — Failure to achieve one or more objectives by an unacceptable margin; significant financial loss to the organization; serious adverse publicity; the project end product would be useless or unusable

- Medium (3) — Clear and measurable adverse effects on scope, schedule, and/or cost
- Low (1) — No loss or negligible scope, schedule, or cost effect

Overall — What is each team member's assessment of the overall level of threat to the project, as informed by Likelihood and Impact?

- High (5) — There is a sense of dread about the risk; they express a high level of concern.
- Medium (3) — There are serious concerns.
- Low (1) — The threat does not seem serious.

A variety of risk assessment approaches are available. The approaches used for any given project should be consistent with the nature of the project and the organization's tolerance for risk exposure. Figure 3.6 illustrates a simple multiplication of the scale values to determine whether a risk is considered high, medium, or low. Ratings are assigned to Risk Likelihood and Impact based on an agreed scale, e.g., 1 (no likelihood) through 5 (virtual certainty).

Ratings are assigned to Risk Impact on an agreed scale, e.g., 1 (minimal impact) through 5 (critical impact — occurrence could lead to failure to achieve one or more project objectives). Note that in the figure, each risk is rated on both its likelihood of occurring and the impact to the project should it occur.

Figure 3.6. Risk Scoring Matrix

RISK SCORE FOR A SPECIFIC RISK					
LIKELIHOOD	RISK SCORE = L x I (Likelihood x Impact)				
5	5	10	15	20	25
4	4	8	12	16	20
3	3	6	9	12	15
2	2	4	6	8	10
1	1	2	3	4	5
	IMPACT ON AN OBJECTIVE (e.g. cost, time, or scope)				
	1	2	3	4	5

Source: Project Management Institute. *A Guide to the Project Management Body of Knowledge: PMBOK Guide*, v1.5. Newtown Square, PA: 2000.

Step three of our five-step process is *developing controls for making decisions that will eliminate or reduce the risk.* In project management-speak, we typically refer to these as risk response plans. Risk response plans should be developed for all risks that are assessed as high overall threats, as well as for some risks that are assessed as moderate threats. The response plan for an individual risk can include any of the following decisions:

- **Accept the risk** — Decide to take no action to prevent the risk from occurring or to recover if it does occur. Be prepared to implement a risk management plan if future assessment of the risk changes.

- **Manage the risk** — Decide to actively manage the risk. In this case, the risk response plan will involve taking one or more of the following actions.

This next step in managing the risk falls into step four of our five-step risk management process — *implement controls to reduce or eliminate the hazard.*

- **Prevent the risk** – Identify actions that address the cause of the risk so you can prevent or reduce the likelihood of the risk's occurrence.
- **Develop contingency plans** – Identify contingency actions that could be taken to mitigate the impact of the risk should it occur. *Example* – A contingency plan for the risk of hereditary illness would be to seek medical treatment. For this contingency plan to be effective, an insurance plan should be purchased before the disease is detected (a contingency readiness action).
- **Evaluate the risk management plan** – Consider the feasibility and cost-effectiveness of each preventive action and contingency plan relative to the project context and to the level of risk acceptable to the organization.

Step five of our five-step process is *supervise and evaluate.* Essentially, this involves putting in controls that allow you to monitor the process and make adjustments along the way. The best way that I've been able to truly monitor risks is to make sure each risk has an associated trigger. Triggers are events, data, or other means of alerting the owner that a risk has occurred. A risk trigger is analogous to an alarm that signals the activation of contingency plans. If the chosen risk response includes developing contingency plans, you must identify a trigger for the risk. For what it's worth, I've often heard from senior executives that the risk management skills of a project manager are the skills valued above all else. They never want it to "hit the wall" before they're aware of it. For risks with high potential impact or for those risks that aren't easily detected, identifying several triggers can be advantageous.

Now I don't want you to expect that just because you've used the steps above to identify and manage risk, your project will be event free. Change will affect your project, and your ability to manage it through project completion is your main objective. If things always went as planned, the world wouldn't need project managers. Therefore, the last step in putting together your project plan and setting your team up to succeed is understanding how you're going to manage change throughout the project's lifecycle.

Manage Change and S.T.O.P. Having Change Manage YOU!

If you think about it, a project manager's job is really based on the art and science of managing change. If one could put together a schedule and execute exactly to plan without any changes, they probably wouldn't need a project manager. A "project change" is any modification to the project that will potentially affect the project schedule, cost, scope, and/or quality.

The most frequent source of change to a project is the change associated with project scope. Project scope is defined as the work that must be done to deliver a product or service with specified features and functions. Project scope is typically finalized after the team has developed the functional specifications or detailed analysis and has made assessments regarding what can get done, how much it will cost to achieve the project scope, and by when it can been delivered.

A project change management process ensures that proposed changes and their associated impacts are assessed and understood *before the changes are implemented.* This provides a common and visible process for maintaining project scope; resolving requested changes; reducing the uncertainty around the existence, state, and outcome of a change that has been requested; and facilitating communication about requested changes among stakeholders.

Typically, a project change management process becomes active once a document has been approved. As it relates to product development, the change management process should become active after the market requirements have been completed. During the analysis of those market requirements, a change control process should be used when you reach an unacceptable threshold against any of your success criteria. For example, if your success criteria are constrained by budget, you would implement a change request when a change occurs that has an impact of greater than 10 percent of the original project budget. This 10 percent threshold, or whatever threshold you're marching toward, would need to be agreed upon upfront by your team and stakeholders.

Any project team member can and should use the change control process to initiate a change. Some typical examples of when a change request should be used are:

- Requests for content changes based on new knowledge of what is possible
- Requests for requirements changes (additions, deletions, modifications, and deferrals)
- Requests for enhancements to the release
- New requirements based on a changing environment
- New requirements discovered during the creation of the technical specification

> *Stop!* Using a change control process isn't meant to be the Wild West. Introducing any change request should be discussed as a team to ensure everyone agrees and that going through the process makes sense! *Think* about the type of change being requested and observe *how* it may be perceived by your entire project team. Once a change request is brought up to the stakeholders, it's hard to take back. Make sure all the proper planning, analysis, and thought has occurred going into the change, and all mitigation steps have been taken.

The following should be considered the Guiding Principles for Change Management:

- **Stop** and document the change before the change is implemented! Most of the benefit of project change control is lost if changes are only documented after the fact. The real power of this process is that it gives management a tool to assess the impacts of a change in advance and to make an informed decision regarding implementation. Project changes often have unexpected negative impacts on the project. A project change control process helps ensure that changes and all their associated consequences are well understood and approved by management in advance. In other words, *stop* surprises to management!
- **Think** about how this "process" is going to be perceived and stress the fact that it is intended to minimize the negative consequences of change. Not only should project outcomes be predictable, they should be predictably better than with an uncontrolled process. You want to add control, not unnecessary overhead. This project change control process must implement control without introducing unnecessary bureaucracy. You should use a one-page request form to record critical summary information throughout the process.
- **Observe** trends and the need to increase the level of control as the project progresses. As the project progresses through successive lifecycle stages, the cost of change dramatically increases and the level of control must also increase. Are there trends that are occurring

or taking shape within the project? A project change control process will provide a historical record, allowing you to answer these questions. Be sure to orient the team around the change. Make informed, collaborative decisions as a team. Change control requires you to research and document the impact of each proposed change, and requires the project core team to collectively approve any decision to implement a change.

- **Plan** to record all project changes, including impact and final disposition. This change management process provides a complete historical record of project changes and their final disposition as a function of time, thereby answering the questions “How did we get here?” and “What can we do better?” The historical record provides a powerful basis for future process improvement.

To truly understand how you’re going to manage the change, you need to understand how members of your team should interact with the process. Figure 3.7 summarizes the responsibilities of key stakeholders in the change management process. Again, note that any project team member can use the request form and the change management process to initiate a change.

Figure 3.7 Change Management Process Summary Roles and Responsibilities

ROLE	RESPONSIBILITIES
Product Manager/ Project Sponsor	▪ Participates as a member of the core team in making and validating final decisions concerning proposed changes ▪ Provides leadership and direction in coping with change impacts, including resource and schedule relief where appropriate ▪ Acts as a liaison to other impacted projects
Project Manager	▪ Administers the project change control process for his/her project ▪ Directs impact assessment activities for proposed changes ▪ Prepares and presents a recommendation for the final disposition for all project change requests ▪ Implements accepted changes by directing the update of all impacted project artifacts appropriately ▪ Maintains a project archive of each request and supporting documents ▪ Communicates the change process and decisions to key stakeholders
Core Team Members (Engineering, Quality, etc.)	▪ Acts as a project change "control board" for all change requests ▪ Participates in impact assessments ▪ Reviews change request recommendations and supporting materials ▪ Make decisions concerning final disposition of change request ▪ Communicate decisions to their respective constituencies
Extended Team and Other Stakeholders	▪ Initiates change requests as appropriate ▪ Contributes to impact assessments as assigned by the project manager ▪ Implements accepted changes as required

Using the S.T.O.P. principles as a framework, Figure 3.8 represents the steps of a project change control process in a process flow diagram.

Figure 3.8 Project Change Control Process Flow

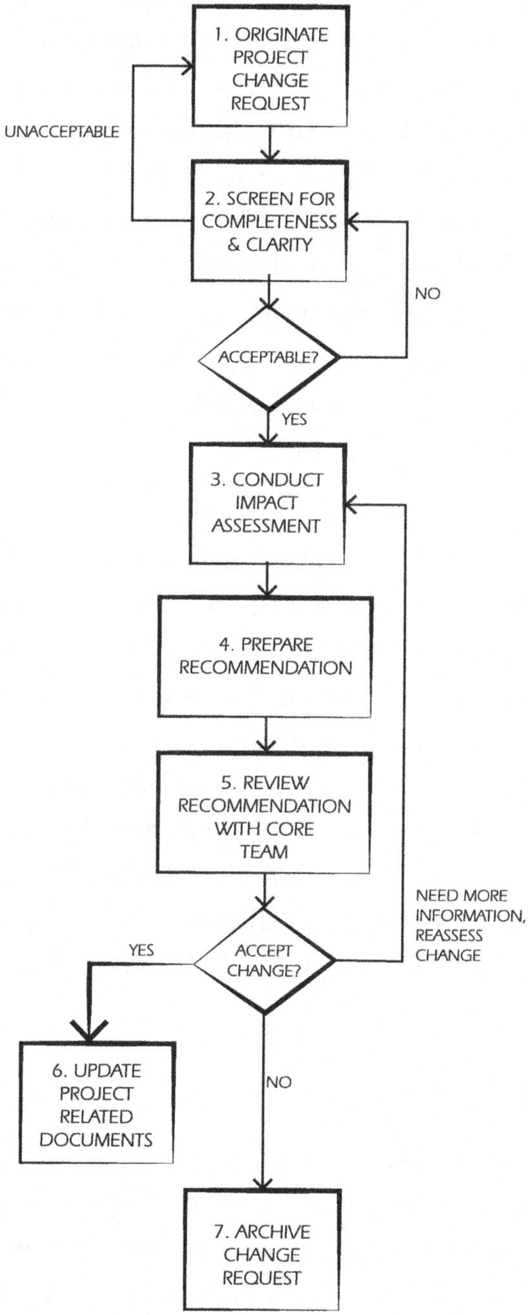

It's important to realize that when change occurs on a project, depending on its size, it affects all that are involved. We see this effect, especially in cases of scope change, when a change in requirements can affect every downstream deliverable the team works on. *When* the change occurs in the overall project lifecycle will determine how the team reacts to the change. Changes early in the project can be perceived as small victories. Not only was the change discovered quickly, but it "saved time" by preventing downstream rework. Changes late in the game will significantly impact the motivation of the team because these changes typically bring increases to timeline and cost — and more work!

As a project manager, it's important to understand this dynamic. A well-planned change still needs to occur and can't be avoided. The cost of not implementing the change could be great if weighed against the business case. Keep the team's motivation high by refocusing them on the project's overall goals and success criteria. Because your project plan is tightly coupled with the market needs the project is meeting, it should be easier to re-enforce its value and benefit to the team. Don't be afraid of the change. Coupling the S.T.O.P. principles with the process above works and will allow you to manage through the situation. Take control and manage the change! Don't let it manage you.

Are You Prepared?

At this point in the book, we've worked through each step of preparing you and your team mentally for your project. This step-by-step process started with establishing success criteria and quantifying project value, was followed by determining the levels of team integration and performance necessary to deliver that value, and led to the different types and levels of communication required for overall project effectiveness. All of these steps culminated in what we called an integrated project plan — a plan geared toward effectively leading your high-performing, cross-functional team in the delivery of a product, and ensuring that the value, scope, and success criteria for a project is well understood by all involved.

By now, if you've followed these guidelines, you've established yourself as the analytical advisor for your project, able to clearly articulate the boundaries of scope, the budget, delivery timelines, and delivery procedure through the product development process. *But are you prepared to lead? Do you really understand project management and when you should stop your project?*

To answer this question, we need to go back to our previous two equations and derive an overall equation for *stopping* a project and for project management performance. In Section 1, we defined value as:

$$\textit{Value} = \textit{Scope} \div (\textit{Schedule} + \textit{Cost})$$

Then, during our discussion of team integration and dynamics in Section 2, we defined team performance as:

$$\textit{Team Performance} = [\textit{Social bonding} + (\textit{Team identification} \times \textit{Team orientation})] \div 2$$

Now, if you consider that communication is roughly 90 percent of what a project manager is supposed to do, you can take this same approach and quantify the amount of communication that is necessary for your project. Use your same rating scale again and, together with your sponsor, assign a value to each section of your communication plan. Assess each type of communication required and its effectiveness. Have you evaluated and communicated project change appropriately? Has your team worked together to identify, discuss, communicate, and manage project risk? You must constantly assess and evaluate communication and ensure that your communication plan is executing properly throughout the project. Once you determine its value, you can derive the ultimate equation for determining whether your project should be *stopped*. Summing the ability to direct and guide a project to maximum value (*Value*), the ability to optimize team integration and efficiency (*Team Performance*), and the overall effectiveness in project communication (*Communication*), and then dividing that sum by 3 to calculate the average, you get the following equation:

$$S.T.O.P. = (Value + Team\ Performance + Communication) \div 3$$

Determining the threshold value of your project will need to be done in collaboration with your sponsor. That value will be significant in signaling when the project should be stopped. Understanding the components making up the equation will allow you to better manage, control, and course correct where appropriate. As my rule of thumb, I would stop a project if this value drops below 75%. However, each project is different and you'll need to determine your own value based on the specifics of the project. If you understand this equation, *you're prepared to lead and to stop!*

The last step in preparing you for your mission in project management is to provide you with a real-life example of how to actually apply the concepts discussed thus far. The story is real, the application of concepts is real, and the results are real. Feeling prepared? Alright; on to Section 4!

S.T.O.P. SURVIVAL TIPS

- **Stop!** As a project manager, your job is to ensure the project is viable. If it's not, make the recommendation to *stop* it. The only way you can do that is by compiling enough data to support your recommendation. Become the analytical advisor for your project by breaking down metrics into the following types and getting the right data your project needs:
    - **S**topping metrics
    - **T**racking metrics
    - **O**bservational metrics
    - **P**rocess/Planning metrics
- **Think** about the type of change control process you need for your project. Project changes can be complex; observe how a change may be perceived by your entire project team. Once a change request is brought up to the stakeholders, it's hard to take back!
- **Think** about how you manage risks. Your team doesn't need another list or spreadsheet to look at, where no action takes place. *Think* about the five steps to risk management and construct the five things you, as a team, need to do to manage risk.
- **Plan** your communication. It's no mistake that 90 percent of what you do is communication. Plan the story you need

to tell by *thinking* about who needs to be communicated to, any constraints to communication, and what outcomes you hope to achieve with the messages you deliver.

- **Plan** your project. A project schedule isn't a project plan! A project schedule is a *tool* used to *manage* a plan. Become a project tailor and create an integrated project plan designed around market/product needs and project outcomes. Do this and you'll turn your plan from one that is inwardly focused to one that ultimately meets customer needs and ensures that all the project work yields the outcomes your organization expects.

SECTION FOUR

Now You're Ready to Proceed

By now you're realizing that the S.T.O.P. principles aren't just applicable to project management — they're a way of life. The principles can be used in just about any situation. We used them to turn our ordinary project plan into a survival plan. When the principles are complemented with the concepts presented in the previous sections, you get a powerful framework and approach for managing any project. The plan we designed isn't just for stopping failure fast — it's for succeeding equally as fast.

I designed Section 4 to demonstrate the value and applicability of the S.T.O.P. principles through a real-life example. Storytelling, to me, is one of the best knowledge-sharing techniques, and it's a great way to pass on factual information. It allows you to find a reflection of yourself in the story... making the material easier to relate to and use.

The sequence and details of the story differ somewhat from the actual events. These changes are by design to keep the focus on the concepts and their value. However, it's important to note that all of the events in the story happened over 12 months. The S.T.O.P. principles and associated concepts aren't meant to take years to implement and realize value. A well-designed "survival" plan will yield results immediately. S.T.O.P. is meant to be used *now!*

An "Everyday" Story

The Problem

The phone rang. It startled Spencer — he was deep in thought, putting the final touches on his project closure document. He looked at the caller ID and saw it was the head of the division, Jack Goldblume, calling. Spencer Carter had been a senior program manager at the organization for five years and thought he had seen just about everything from a project management perspective. Intuitively, he knew picking up his phone wasn't going to result in a casual conversation — this call was going to be *different*.

Lately, the scuttlebutt around the office was about one of the more mature product lines and the quality issues it was experiencing. It had been rumored that customer complaints had made it all the way up to the CEO, Mike Deegan; not a firestorm anyone wanted to be a part of.

Spencer paused, took a deep breath, and picked up the phone. " Spencer Carter..."

A gravelly voice responded, "Hey Spencer, it's Jack. How are ya?" Without waiting for a response, Jack half-jokingly put the real purpose for his call in play, asking, "Ready for your next assignment?"

Spencer knew the small talk was all just a set up, a necessary evil to unraveling the true purpose of Jack's call. "Great Jack," he responded, "just dotting the i's and crossing the t's on this one. It'll be good to put it to bed. What assignment do you have in mind?"

"Well Spencer," Jack said, "I'm not sure if you've heard about the issues regarding product Nova. The number of product defects is growing by the day and customers are now calling Mike Deegan directly. I know the ink's not dry on your previous project, but are you up for a challenge?"

It seemed like the gears in Spencer's mind were audibly clicking. He felt an increasing sense of unease. Not sure how to answer, Spencer hesitated, then replied, "I guess so. Can you tell me any more about what you'd like me to do?"

Spencer could almost see Jack's smile on the other end of the line as Jack responded, "Spencer, we want you to go in and fix it!"

Feeling a bit like a fly caught in a spider's web, Spencer's voice raised slightly, "Fix it sir?"

"Yes, *fix it* Spencer," Jack said. "Listen, I need to run. Mike Deegan should be here any minute. I'll let him know you're going to resolve this issue. Get in touch with Carl over on product Nova — he can give you the details. Spencer, I really appreciate this. *I'm counting on you.* If you need my help, just let me know. Deegan's here. Gotta run. Enjoy the rest of your day."

Then Click; the connection ended and that was it. In a matter of 30 seconds, Spencer's next 12 months was fated — and it would challenge all of his abilities as a project manager.

Before calling Carl, the head of product management for Nova, Spencer did a bit of asking around the office to determine what people knew regarding product Nova. The word was the internal support staff had become increasingly agitated by the growing external user complaints. And the sales team was screaming that, due to the issues, they would never hit their sales targets.

Management had lost trust and confidence in the team's ability to execute. No confidence and trust resulted in no product funding. Basically, "Go fix the issues you have, but we're not investing anymore in you because you'll just cost us more money", is how the project team interpreted management's decision. The lack of further investment caused their customer base to be extremely dissatisfied with the team's perceived lack of innovation. All the work that went into each release didn't seem to make a difference. The customers felt that the product was stale and the competition was starting to sound very attractive. It was a death spiral.

> Information gathering is a key first step to stopping. Get as much context as you can about the situation before you enter it. Once you're in "it," it's too late. Take the time to stop, gather information, and allow yourself to form the approach and next steps you may need to take.

Adding further pressure to the situation, the team had made external commitments to the market, with the intent of delivering new releases quarterly (if not sooner). Enhancement requests were starting to pile up with no strategy for addressing them. The team tried to rally around the situation, but because they were located all over the world, it was hard to coordinate their efforts, bond socially, and build team relationships. All this frustration resulted in a team reacting to firestorms, never appearing to make progress or having time to course correct, and operating without a plan. The need for a *survival* plan was definitely required!

Stopping to Identify and Assess Hazards

Success Criteria and Assimilation — An Inside-Out Approach

Spencer set time up on Carl's calendar to talk. Carl had been with the product for roughly five years and had a lot of experience working with his product management team on the East Coast and the development team in India.

Carl started the meeting by welcoming Spencer to the project, "Hey Spencer, welcome to the team! Jack told me you'd be calling. How are you?"

"I'm good Carl. Can you tell me a little bit of what's goin' on over there?" Spencer asked.

"We're knee-deep in quality issues and senior management is breathing down my neck. Imagine trying to fix a plane while it's still flying... that's about the best way to describe what's going on. Not sure what you can do, but I am glad you'll be working with us to sort this out and get back on track," Carl replied.

"Thanks Carl," Spencer replied. "I'm not sure either. Would you mind if I started meeting with the team to get a better sense of the situation?"

"Of course not, do what you need to do," Carl responded. "They're all extremely busy fighting this blaze, so good luck finding time on their calendars. If you run into a problem, let me know."

Joining the team in mid-flight wasn't going to be easy. To be honest, Spencer still didn't know what the "project" was.

He was given the assignment to "go in and fix it." Now he just had to figure out what "it" was; and "fixing" it was going to have to happen in the air while this plane was flying because, per Carl, "There was no tolerance for *stopping* anything!"

The last thing this team needed was an outsider coming in and thinking he was going to "save the day." Constructing success criteria had to be done together, and his only chance for assimilating into the team was to get them to trust him by conveying that he truly understood their problem – which was now "*our*" problem. They had already been through a lot, and he wanted to give them the opportunity to express their frustrations and "get it all out" before he tried to jump in and find a solution.

The first step was really a listening exercise. Spencer started his situational analysis by meeting individually with each team member to discuss his assignment and what their shared goals and success criteria should be. The majority of the conversations were done over lunch and ad hoc. He couldn't interrupt their flow, so he had to look for opportunities to generate the conversation. This approach afforded him the opportunity to socially bond and start building relationships while gaining valuable information (and about five pounds).

> Performing a situational analysis is one approach for temporarily stopping the project, without anyone perceiving that it's been stopped. Getting people to talk allows them to slow down and think about their situation. Giving them time to think and hear themselves describe the problem out loud creates a more constructive environment for problem-solving.

They all quickly concluded that their problem wasn't just about product defects. It was far more complex and it involved setting appropriate expectations, delivering value, and improving trust at all levels. Spencer knew that the team would essentially have to work inside out to solve this problem. To do that, they'd have to agree on what the ideal goal and end state was and then work backwards. They would have to develop a plan that worked internally toward that external goal because the only way to make sure their goals were accurate was to validate them based on who was going to ultimately benefit from the outcomes.

Starting with their customers as the ultimate beneficiary, the team realized the only way they were going to change their perception of the product was to fix the defects *and* innovate — a truly tricky situation. Happy customers equal more revenue. More revenue equals more product investment. Dissatisfied customers equals market share erosion and a rapid, self-fulfilling death spiral. Customers wanted to see change and the team knew they had to grow their way out of this problem. Unfortunately, to do that, they had to convince management to re-invest.

The first step in regaining management's trust was for the team to communicate the root cause of the issues. They had to convince their internal constituents that they really understood the problem and were committed to and capable of solving it. Spencer knew what this team needed was a public relations campaign — and they needed it now. He immediately began constructing the communication plan. His stakeholder analysis resulted in both internal and external stakeholders — the senior management responsible for the product line and the client community using it. The stakeholder analysis also indicated that the team's governance structure was lacking.

To get the executives to help solve the problem, Spencer organized them into a steering committee. Once formed, he immediately got the members together to discuss roles and responsibilities — what the team needed from them and what they needed from the team. Collectively, Spencer and the steering committee agreed that the SC would run monthly meetings where project team members would give progress reports and identify issues needing management's help to resolve. His plan was to use these meetings as the initial step in the communication plan. He knew that if he could get the SC excited about the progress, he could use them as additional advocates for the project team's message.

The first meeting was extremely successful. Spencer stated the team's mission, the root cause of the problems, and their plan of attack. After seeing how pleased the SC was with this initial approach, Spencer took a chance and asked, "If we can show incremental success on the execution of this plan, can we use the upcoming all-staff meeting to communicate our results?" The extent of the damage around the product line was pretty well known and Spencer knew they needed a broad-reaching communication vehicle to showcase their work and get the word out that the team was turning this ship around. The recommendation was well received and the SC agreed, saying, "If the team has results to show, then we want you to show them!" Spencer knew that if the team was successful, management would be viewed as successful — *an inside-out approach.*

The team's key messages had to be positive right out of the gate, focusing on big wins and using impact words that the executives and sales team members would understand and gravitate to. Enlisting the internal marketing department was exactly the help Spencer needed to accomplish that goal. Together the team constructed the message, the vehicles for delivering the message, and the timeline for its delivery. The communication plan was taking shape. Now the team had to focus on execution!

Many times in project management, to get the inputs and information you need from the program level, you almost have to take an inside-out approach. What I mean by that is you can't put the current project on hold while you go and figure out the program-level information. Although this approach makes sense, it won't be tolerated by senior management. Part of project management, especially within technology product development, is fixing the plane while it's flying. You have to keep the project moving while you collect data and ensure that the program-level information is intact and in sync with the goals and expectations of the project. For example, if at the program level there are goals and objectives around growing the product line or business by x percent, but at the project level there are no growth-type projects, there could be a problem. Having the opportunity to ask those questions upfront is great. However, many times you won't have that opportunity. Things are already moving, commitments have already been made, and you are coming into the project in mid-flight. Can you yell *Stop!* here and get away with it? Not until you have the data you need. Then you can yell *Stop!*

Further Observing to Assess Additional Hazards

The front door closed tightly with a reassuring thud as Spencer arrived home after a long day at work. Even though he was home, he couldn't stop thinking about the current situation. He knew executing against a plan wasn't going to be as straightforward as management thought; there's always more to solving a complex problem than "just fix it."

He had to take a step back and observe the environment in which this team was currently executing. How were they organized around the work they did? Who was working on what? Were they dedicated or did they get pulled from working on functionality to be put on the issue response team if a high priority bug was reported? Spencer knew he couldn't answer these questions, so he made a dedicated effort to push them out of his mind and enjoy the rest of his evening.

The next day Spencer awoke with a plan to get to the bottom of some of the questions rattling around in his head. His first step was to set some time up with the engineering director, Robert, to get his perspective. Robert was full of energy and fire. His inspiration came from the innovative technology companies out there who were in the headlines, making a name for themselves. Although full of spunk, Robert was also extremely frustrated that this product line he worked on could never seem to get "there"— a place to actually *innovate*.

Knocking on Robert's door, Spencer announced himself. "Hey Robert, how's it goin'? I was wondering if you have a couple minutes to chat?"

"Sure Spencer, what's on your mind?" Robert replied.

"Robert, I'm interested in learning more about how the current core team is organized. Is everyone getting along and do they all know what the others are doing?" Spencer asked.

Robert paused for a second and then began sharing a bit of history Spencer wasn't expecting. "Spencer, it's been a rough ride. My guys are getting whipped around from one issue to the next and we barely have enough time to get the context for what we're actually trying to do. Product management always seems to be changing requirements at the last minute. I'm not convinced they even know what they want. And to top it off my team feels like they're coding in the dark."

Spencer was a bit confused by Robert's last comment. "I'm sorry Robert, I don't understand. What do you mean they feel like they're coding in the dark?"

Robert went on to explain that his team was not the original owner of the system for which they were now wholly responsible. The original owners were long gone due to continuous organizational reshuffling. That meant that this team, from product management to development to quality assurance, really didn't have a truly detailed grasp of how this product behaved, especially when introducing new features/functionality. They would fix one issue and break another. They could never seem to dig themselves out of their own hole. Robert's perspective was helpful, but it wasn't complete. To construct a kite structure core team, Spencer knew he had to get the product management perspective of the situation as well.

Jillian was the product manager reporting to Carl. She had been with the product for years and had great depth of knowledge concerning product features and customer perspective. Later that day, Spencer caught Jillian on her way to lunch and asked if he could join her. She politely agreed and the two headed down the elevator to the local dine-and-dash.

"Jillian, thanks for taking the time to meet with me. I hope I'm not interrupting any plans you had," Spencer said.

"No, Spencer, not at all," she replied. "The company would be great."

After a few more pleasantries, Spencer dove in by asking Jillian about the quarterly release schedule and the customer requirements.

Jillian replied, "Spencer, we are victims of our own self doing. We get about five minutes to construct a roadmap that we have to live with for the entire year. These roadmaps basically get turned into customer commitments. We do our best, but we never seem to have enough time to get it all done. Given the scale of the current problems, three months is just not enough time to get a release out the door. Coupled with the quality issues we're having and development breaking stuff all the time, we constantly find ourselves in an uphill battle."

"What happens when we don't deliver? Do the customers get upset?" Spencer asked.

"Well, some do," Jillian answered. "Honestly, as an organization we continuously mismanage their expectations. We continually shoot ourselves in the foot, a self-inflicted wound. Compounding things, it pushes off the other things *they really wanted us to get done.*"

"Jillian, did you say that you basically start each release at the end of the last release?"

"No, oh no," she replied. "We try to get a head start. Most of the time we're working on three releases at once. Kathy, our project manager, is great. She keeps us going and on task. We would die without her."

Spencer almost choked on his teriyaki chicken. The picture was becoming a bit clearer in terms of true root cause and potential next steps. Returning to the office, Spencer gently closed the office door behind him and looked for some quiet space to stop and gather his thoughts. He was confused about why the team committed to delivering functionality without planning how they would actually deliver it. In addition, he was dumbfounded on how a project schedule could actually get created for all this work.

Without hesitating, Spencer picked up the phone and called Kathy, the project manager located on the West Coast. "Kathy, it's Spencer. Long time no talk." Kathy and Spencer had a bit of history earlier in both their careers, reporting to the same manager.

"Spencer, good to hear your voice," Kathy said. "Have you talked with Sinclair lately?"

"No, not really," he responded. "I need to catch up with him to see what he's got going on. He's an interesting guy."

Spencer quickly ended the small talk and got right to the point. "Kathy, I've been talking with the Nova team and I just don't understand how you're even managing this effort. Can you show me your project plan and schedule?"

Kathy quickly set up a web meeting and walked Spencer through her documents. It seemed that her project plan had the right sections filled out, but context was missing. There wasn't that analytical information he was looking for to back up the success criteria she had listed. The risk items were the same for each release. Spencer saw tight timelines and constrained team bandwidth, both of which could cause project timelines to slip.

"Kathy," Spencer said, "Would you be up to exploring a different way to manage this team?"

He could hear her grinning over the earpiece, "YES! Please, let's set some time up to figure out a better way," she replied.

With all these challenges, where did Spencer start? Well, the first thing Spencer did was he tried to get the team to just *stop* - stop doing, stop finger pointing, just plain stop and take a breath. Next, he wanted them to throw away any and all project schedules they had created. Not actually throw them away in the garbage. Instead, keep working, but build the project schedule at the program level and work their way down to the projects. He wanted them to really *think* and focus on the information *in* the deliverables, and not just the deliverables themselves. Finally, he wanted the entire team to *observe* and understand the priorities of the business: What was driving this product from a business perspective and what did they need to do to meet those priorities? Once they could obtain clarity, then they could start to construct a plan. He then jotted the following on his whiteboard to help the team deconstruct the problem and determine the right set of objectives for a solution:

Value = Scope ÷ (Schedule + Cost)

He determined that the scope of what they were delivering was relatively low. Cost to deliver was high because essentially this was all re-work — fixing issues that made it to production. Schedule was high as well because enhancements were pushed off and late. His ideal state was to have a value equation at 100 percent. So, keeping his scale consistent at 1–10, he assigned values for the ideal state and the current state:

Ideal *Value = 10 ÷ (5 + 5) = 1 or 100%*

Current *Value = 5 ÷ (8 + 10) = 0.28 or 28%*

He assigned schedule an 8 indicating that the schedule component of the project was operating at roughly 160 percent of its ideal state. He took the same approach with cost. Because project costs were roughly twice as high as the team originally had forecasted, due to all the rework, he assigned a 10.

Spencer concluded that the team was currently delivering roughly one-third of value relative to their objectives and product line goals. More scope and timely delivery, at less cost and less post-production quality issues were needed. Spencer, working with the team, came away with three main objectives. These objectives would define what "it" was to Spencer and "what" the team would need to do to be successful. After they figured out the "what," they could figure out the "how." The three main objectives were:

1. Fix the high-priority defects and reduce the overall defect list to something manageable. (Benefit: reduce friction with the current customer base.)

2. Increase quality by better understanding the overall system. The idea was not to recreate documentation for documentation's sake, but to create a pre-step geared toward a baseline understanding of what the system currently did and assess the impact of any further design extensions in that specific area. (Benefit: increase the team's effectiveness.)

3. Introduce new functionality that the market and sales teams had both agreed were needed. Focus on any items that were long overdue. They didn't have to be big; they just had to have impact. Then communicate, communicate, communicate the benefits and value the team was delivering. (Benefit: demonstrate responsiveness and increase trust and confidence to both the internal stakeholders and external customers — further increasing the likelihood of additional investment.)

These priorities defined the "what." Now Spencer had to figure out the "how." He started to construct equations associated with relative value to determine where he needed to focus and prioritize his efforts. He wrote the following equation on his white board and jotted notes as he applied a value to each variable:

Team Performance = [Social bonding + (Team identification x Team orientation)] ÷ 2

He wanted to keep things simple and consistent so he gave the value of 1 a meaning of perfect state. He would then multiply the entire equation by 100 to convert it to a percent.

- Social bonding = 0.8. A distributed team of this size would never be at a 1. The opportunity for social bonding is just too limited.
- Team identification = 0.8. Again, the distributed nature of this team would only allow team identity and autonomy to be a 0.8.
- Team Orientation = 1. The team should always be oriented around the work.

Team Performance = (0.8 + (0.8 x 1)) ÷ 2 x 100 = 80%

He then wrote down how the team was currently operating:

- Social Bonding = 0.4. Team had very little opportunity to socially bond. The only social bonding that was occurring was discussing the product's current issues.
- Team Identification = 0.4. Because of the product issues, this team was not allowed to have any autonomy. Management had lost trust and confidence in them and was essentially micromanaging their every move.
- Team Orientation = 0.6. The team was oriented around the work in front of them, but was not oriented around delivering overall value or growing the product.

Team Performance = (0.4 + (0.4 x 0.6)) ÷ 2 x 100 = 32%

He concluded that this team was NOT operating efficiently relative to the overall objectives. They were operating at approximately one-third of the efficiency they should be. They needed to stop how they were organized and re-evaluate their approach for delivery.

Developing and Implementing Controls — Making Decisions

A Project Manager Becomes a Program Tailor

Spencer threw down his pen as he marked up the release project schedules. "There's got to be a better way to do this!" he muttered. "I've got to figure out how to maximize the value of what we deliver and increase the team's efficiency. And I've got to do this while improving overall communication and understanding."

He took his assumptions and calculations on team efficiency and value and combined them with a value for communication, considering the lack of any previous communication plan for this product line, and concluded, "When adding up each of these values, a current project should be running anywhere from 75 to 100 percent. If it runs below 75 percent, we have to stop the project and apply mechanisms to each variable to determine if we can recover the project." It was then that he coined the S.T.O.P. equation:

S.T.O.P. = (Value + Team Performance + Communication) ÷ 3

Spencer wasn't ready to quit, but he was ready to *stop!* He began categorizing the work based on work type. He theorized that they'd need different resources for different activities. Fixing defects/bugs had different resource needs than adding new enhancements.

Spencer put on his "project tailor" hat on and got to work. He first recommended splitting the work between defects and releases. He wanted a clear line of sight based on the different types of work required. To do this splitting, he adopted a concept from Agile development — creating a "backlog" of all defects, ranked by priority, that was estimated and iterated in two-week intervals. The team would review this backlog every other week to make sure the priorities still made sense and the team was working on the right items.

Working with Robert, he made sure that the resources working on this backlog were separate from the resources working on the release. The defect resolution team would need to have its own identity relative to the new enhancement/release team. Team autonomy was the only way Spencer was going to achieve that goal.

Second, based upon business priorities, the team re-evaluated new functionality commitments. Spencer wanted to see the list of features mapped to places in the system where the team had little context or understanding of how the current code actually worked. He wanted to be sure the team knew what they were getting into, and planned for it appropriately, regarding any new feature that was being introduced to the system. Anything new had to have documentation describing the feature both from a market value and technical standpoint. In addition, any new feature was required to have a robust test set associated with it, as well as a regression test set to ensure they weren't breaking anything by introducing something new.

Finally, he wanted the team to build a plan and schedule that tied all this together. The plan had to be based on the work needed to complete the items, NOT the end-of-quarter delivery date. It was critical for the team to understand what they knew and what they didn't know before setting any expectations on delivery. If they couldn't communicate the value and describe what "it" was, then it needed to be explained further. Basically, it wasn't allowed to leave the conceive phase until all involved could consistently describe what "it" was.

Once the entire cross-functional team agreed on the value and goals of the new functionality, the items were grouped into feature sets in what they called "work streams", as seen in Figure 4.1. Each work stream was then deconstructed to explain all the work that went into producing it. Only after analysis was complete and the rollout and launch plans were constructed was the team allowed to put the item into the release queue and schedule. This framework would allow the team to work in parallel on multiple items, but only release features that were truly ready for production. Some features were more complex and were naturally going to take longer. That was fine, but there was no reason to wait for the smaller features — where the team knew exactly what needed to be done — to be complete before moving on to others.

Figure 4.1 Work Stream Schedule

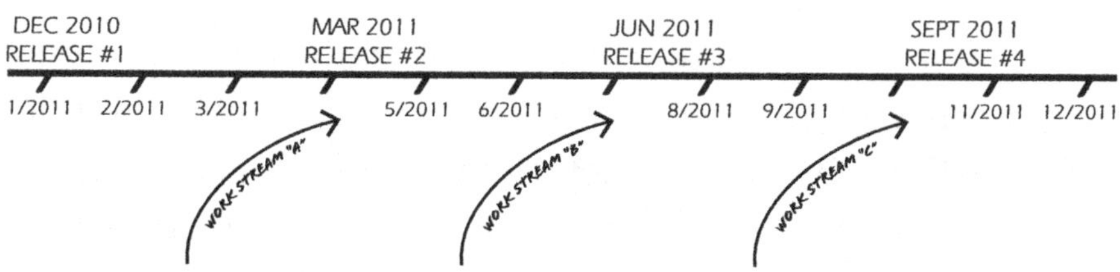

Spencer and Kathy knew that each work stream had to be kept independent of the others because each one had different constraints and levels of understanding. Work stream A, for example, may have needed much more market research relative to features in work stream B. Spencer explained that to the team by saying, "So, let's take that all into consideration and track that in the schedule accordingly. Don't force-fit it in. It will never get done on time. *Think* about the work required and plan it from there, then determine what release it fits into."

This approach allowed the team better visibility into what they could and couldn't commit to delivering. It also gave them the flexibility to deal with some of the fuzzy gray area that typically exists in the front end of the project, without being completely strained by a waterfall/serial approach to development. They soon realized they could better commit to new features being delivered with higher levels of confidence – which is what sales and the customers wanted. In addition, the work stream schedule prevented them from over committing on things they weren't sure about. Spencer explained further to the team, "Allow the level of information we have to dictate our confidence as we progress through each deliverable in the work stream. If we can do this, we can succeed! Deliver value; deliver quality."

Spencer knew this macro approach to building the schedule was initially going to be perceived as a radical change. But he worked with Kathy to construct the message behind it. He had to sell it to the team as a small, incremental, change to how they did work. The only radical change was their focus. The team's focus had to be on value. Each release would have to be weighed by value. If enough value was in a release, then it could be released into production. If a release wouldn't have an impact or enough value, then the team would *stop* and not release.

T-Minus and Counting...

The next challenge Spencer faced was dealing with the overall constraint of delivering quarterly updates/releases to the market. To do that, Spencer pulled a trick out of his hat from his biotechnology days when he used an XP approach for implementing and validating new quality systems. Spencer created what he called a high-level T-minus schedule – separate from the work stream schedule. This T-minus schedule tracked backwards from the time the release needed to be delivered to when the market definition needed to be complete, and all major deliverables in between.

By then overlaying the two schedules, the team could determine what was required for a feature to be eligible for a release based on the work, NOT on artificially committing to dates or trying to force-fit it into a quarterly release. Once a feature had met its requirements for the release, it was removed from the work stream schedule and put into the release-specific schedule.

This multi-level approach would increase the team's line of sight dramatically because they would be able to better predict the characteristics of quarterly releases and set expectations appropriately across the entire year. Spencer explained to the team that the real outcome of this approach would result in their ability to increase the quality and scope within each release. They could literally do more by doing less.

> Developing the process of the work stream schedule and T-minus schedule is another way to stop the line. Build "stop" into the process.

The team did realize that, yes, at times, they would have to commit to certain features that they didn't have as much confidence in as they would've liked — that's the nature of business. But at least now they had visibility and transparency into forecasting and determining the tradeoffs that needed to occur, well in advance. And this made communicating internally to senior leadership and externally to the market much easier. There was now more meaning and context in how to use the tradeoff matrix Kathy had in her original project plan. It was a compass directing the team to optimum value and stopping the work when value was diminished.

From Fuzzy to Clear

A whistling sound could almost be heard reverberating through the office. "Robert is that you? I haven't seen you this happy in years." Spencer said.

"Yeah, all this activity lately has got me energized. I'm really excited!" Robert replied. "Hey I've been reading these articles about how all these Silicon Valley tech companies innovate. I can't stop thinking about why we can't do something like that. We seem so tied to our current process of review points. It just doesn't seem like it leads to quick innovation."

Spencer began to smile inwardly. "There's something you're thinking about Robert... I can see it written all over your face. What did you have in mind?" he said.

"Well, we're starting to get an idea of what the top priorities are across our product. But we still need more detail and I just don't know why we need to wait. Why can't we start working on these now?" Robert asked.

Spencer knew that the biggest challenge he and Kathy were going to have, from a project management standpoint, was trying to become more actively involved in helping the team clear up the fuzzy front end of development. For the most part, the team understood when features were ready to be tracked on the work stream schedule. But understanding how less-clear ideas were allowed to mature into valued feature sets was still somewhat of a riddle. To Spencer, Robert's question almost seemed like an invitation too good to be true.

"From my standpoint, our biggest challenge is figuring out what we don't know. What do your technology gurus say about that?" Spencer asked, almost prodding Robert at this point.

Robert quickly brought up an article about rapid innovation and prototyping and showed it to Spencer. "Spencer, if we could do this, we could really start to innovate and solve some of these product issues at the system level, not just at the local level."

The fundamental idea of the approach was to allow the team time to explore ideas and problems that needed solving *before* constructing any market requirements.

Robert continued, "What we need to do is essentially allow the problems and problem statements to be better understood. Give the team time to build context and clarity."

Spencer called Carl into the office to bring him up to speed on the conversation."Carl, Robert and I think he's just discovered the fountain of youth for your product. You may want to hear this." An energized Robert then explained the concept to Carl.

"So what you're telling me is that you want to give the team small increments of time to think about those hard topics plaguing our product. The issues that are always "too big" to address – so they never get addressed. Right?" Carl said.

"Yes sir," Robert responded.

Spencer quickly added, "This process has to be small and quick to appear not to disrupt or distract from the current release work that's going on but we need to allow these issues to be better understood. And once an idea is understood well enough..."

Carl quickly jumped in, "...we force its prioritization based on the other items in the queue. Is that right?" Spencer and Robert hadn't even thought it through to that point, but absolutely loved where Carl took it.

Spencer quickly responded, "You got it, Carl. What Robert is telling you is this approach prevents the product plan from getting old. We incrementally make progress on those issues that were always too complex to understand and tackle. We continuously innovate, solving problems, and constantly challenge our current product roadmap by ensuring the highest priority items with the highest likelihood of success are the ones getting into the release."

> If you allow your team small increments of time to conceptualize solutions, and complement them with incentives to bring those solutions to market, you have a recipe for continuous improvement using the creative disruption techniques that yield value to both the internal team and the customer.

Figure 4.2 The Fuzzy-to-Clear Delivery Process

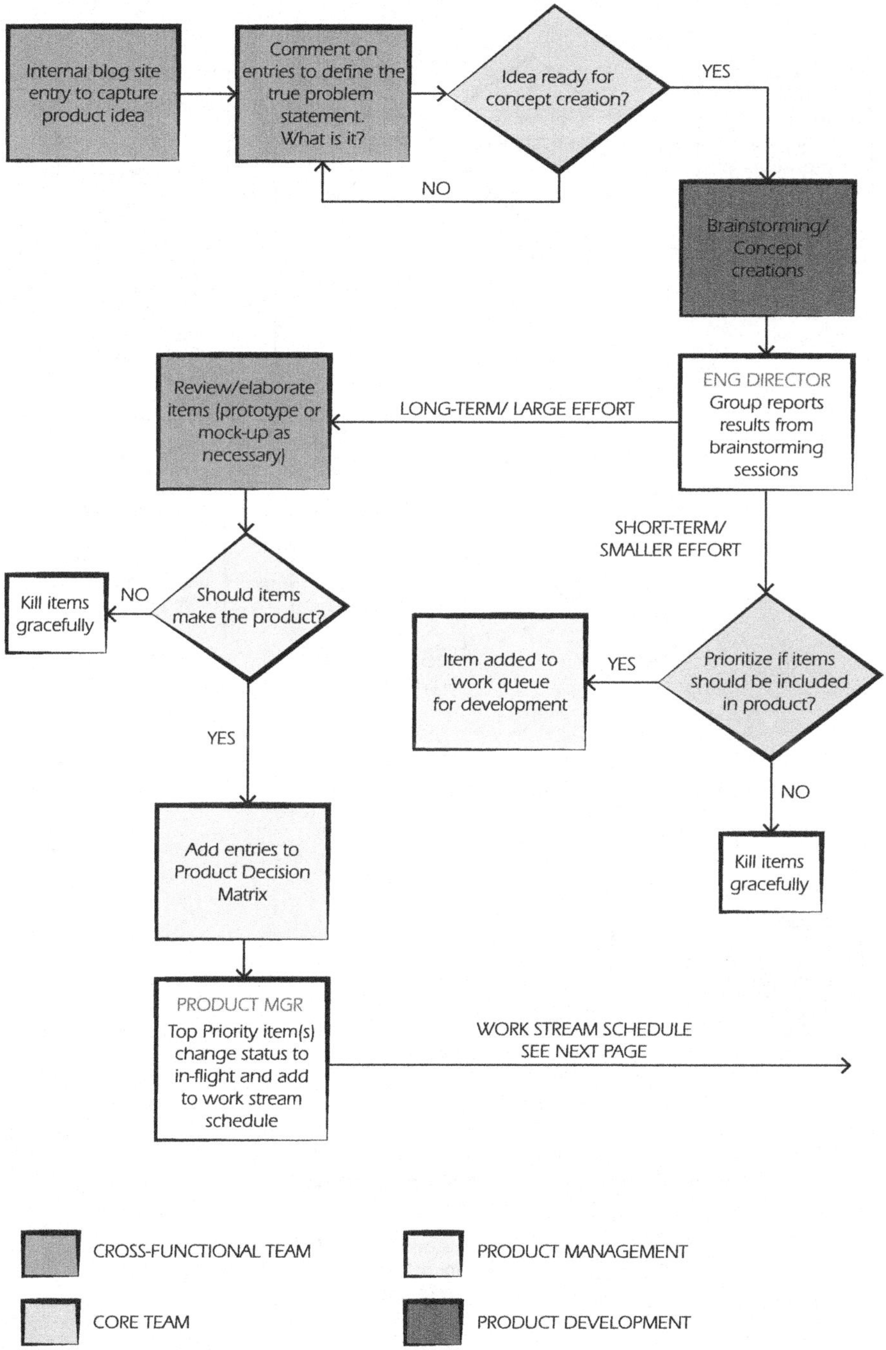

Figure 4.2 (continued) Work Stream Schedule

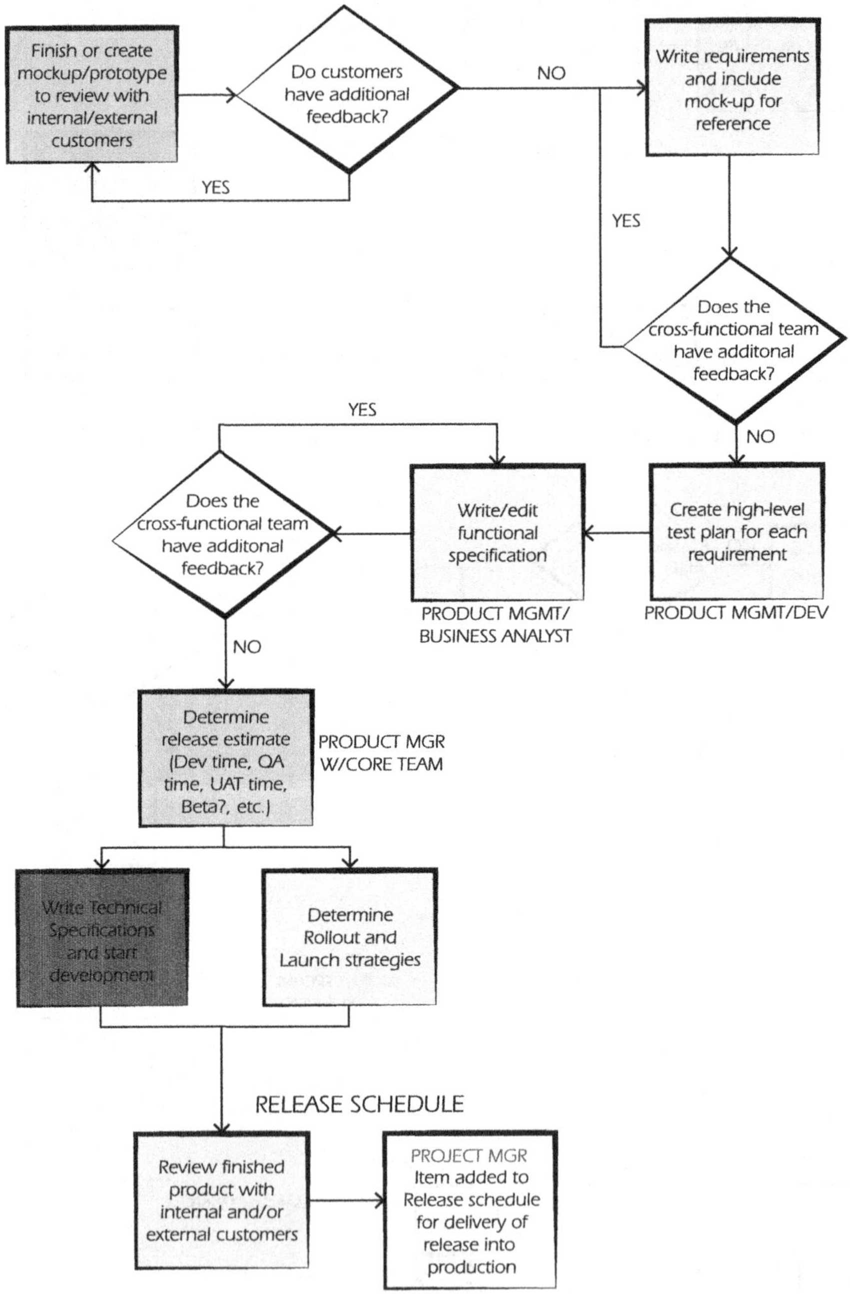

Carl's pause was so long Spencer and Robert checked to see if he was still breathing. Then Carl responded, "So, this process is a bit *disruptive*, huh? I like it!"

Spencer and Robert ran back to the whiteboard to further construct the plan to "innovate." They kept Carl in the loop to talk further about market research activities and what else needed to get fleshed out when constructing a problem statement. Together all three constructed a plan based on people, process, and technology — applying all three appropriately to clear up the fuzziness of innovation and gain both internal and external consensus along the way. Figure 4.2 describes this process.

The team adopted scrum concepts, meeting every other day and using sync meetings to gather with the cross-functional team every week. There were no more excuses for someone "not knowing." They were in this together!

The team was on the same page and that page was written from a plan — a plan created and managed by the project manager. The team was now *truly acting as a team*, speaking with one voice, working toward shared objectives, and all working from the same data points.

As the team began to execute and see results from using the work stream and T-minus schedules, Spencer could sense their excitement. This small disruption to how they accomplished their work was a pressure relief valve as the team realized the approach increased their chances of delivering value and achieving success.

So was this process iterative, serial, Lean, spiral, or waterfall? Well, probably all of the above. But I'm not sure anyone in senior management cared. What they did care about was that the team now had project schedules that were designed around business priorities, success criteria, and allocation/work effort realities.

Supervise and Evaluate

Back to the Team Objectives and Back to the Plan

The car door shut. Spencer turned the ignition, rolled down the windows, cranked the stereo, and enjoyed the small victory as he drove to work. Together, as a team, they had successfully implemented change and the change was in process and working. The latest tracking metrics scorecard showed the team delivered two releases using the new process and reduced the number of patches and client calls after the release. Bug counts were down and confidence was up. Their successes had been communicated to management and functional leaders through the SC process as well as at all staff meetings.

As he walked into the office, Spencer could feel the energy coming back to the team when Robert stated, "Sure is nice now that the pressure has eased a bit."

Spencer responded, "Yeah, it's nice. Nice to have a little breathing space and a bit of autonomy, but we're not done."

Pointing to the list of objectives posted on the wall, Spencer exclaimed, "We really only did the first one of these. Number two is ongoing... We still have to gain investment dollars if we're going to innovate and change our customer's perception of us."

Overhearing the conversation, Carl interrupted, "This product has the potential to be great. We just need to gain the executives' trust so they're willing to invest in us." They all agreed that the last objective was getting investment dollars. They needed to enhance functionality and make an impact with the customer base.

> Supervising and evaluating are another way of saying "monitoring and controlling." Make sure the plan you have in place is working. If it's not, stop and re-evaluate.

Spencer went to the whiteboard, saying, "The last time I talked with Jack, he said something about an investment meeting in three months."

Carl quickly responded, "Yeah, it's a part of our annual cycle. It's when all the product lines basically compete for investment dollars. This is the first meeting of several, where we marry the strategy of the portfolio with the direction of the product lines."

"Then that's the meeting we've got to make an impression at," Spencer replied. I'm not sure how things worked in the past because they never invite me to those meetings, but we've got to show them this year's different. If this is a competition for dollars, we need to get as much of the pie as possible."

Robert was beaming. "But how are we going to determine our product strategy in three months?" he said.

Carl replied, "We have to give them context. Show them not what we're doing for next year, but for the next *three* years. Make them realize the decisions they're making for next year will impact the growth of the product for the next three."

Robert, still a bit perplexed, said, “But to do that, we need more details around ‘what’ we want to do. How can we get there?”

Spencer quickly got in between the two, “Listen, the list of enhancement and customer requests has got to be huge, right?” Both nodded in agreement. “Then let’s categorize this work and get focus teams to determine the top three in each category. We’ll then combine the lists, create our three-year product strategy and, based on the priority, our tactical plan for next year.”

All agreed that the time to do all this was going to be tight, but if they could pull it off, it was likely to be well received by the funding committee. Spencer took the lead and created a rough timeline with intermediate milestones of when the work needed to be done. The three then set off to pull cross-functional team members together to kickoff the focus teams, provide direction, and get them working.

As the weeks passed, the focus teams made progress and the three-year product strategy began to take shape. Communication between the functions was at an all-time high. Spencer helped Carl consolidate all the information from the teams so that it was easier to analyze. It was clear the deficiencies in the product weren’t just small and local, they were systemic. However, addressing the deficiencies meant that they could grow the product. This wasn’t just a story of sustaining the current customer base, but a story of growth. He ensured that Robert and Carl were on track to take the information and construct the presentation for the funding committee to include high-level estimates for cost and benefit. The message was clear and concise: "Invest now and grow tomorrow!"

Spencer knew that fixing "it" was getting closer and closer to being realized. He had done everything he could to complete the project objectives. They had improved quality, incorporated sustainable change, and created cross-functional communication that was timely and efficient. The program now had a plan to not only survive, but to thrive. It was now up to Carl to sell it and the funding committee to invest in it. Spencer went back to his equations to validate the progress. First he evaluated the team's performance by writing the following on his whiteboard:

Team Performance = [Social bonding + (Team identification x Team orientation)] ÷ 2

- Social Bonding = 0.7. Due to the success and the creative disruption that had been implemented, social bonding dramatically improved from a 0.4 to 0.7. The team still needed time to advance to the ideal value of 0.8, but progress had been made.
- Team Identification = 0.7. The team went from not having an identity to one of innovation and growth. Their delivery process had been recently showcased to the greater organization and teams were wondering how they could standardize the process. A huge improvement from 0.4 to 0.7. Given time, they would be at a 0.8.
- Team Orientation = 0.9. The team was oriented around value. They had developed a strategic plan to deliver that value and were organized to execute.

Team Performance = (0.8 + (0.8 x 1)) ÷ 2 x 100 = 80%
(Ideal)Team Performance = (0.7 + (0.7 x 0.9)) ÷ 2 x 100 = 67% (Improved)

His conclusion was that the team had increased their efficiency from 32 to 67 percent. They now had the right context and understanding of how to work with each other and deliver results.

Spencer next looked at overall value:

Value = Scope ÷ (Schedule + Cost)

He determined that the scope of what they were delivering was now relatively high. They hadn't released any major enhancements, but they had been extremely successful at small incremental changes to the product. Patches after production had decreased and, as a result, total maintenance costs had also decreased. Because they weren't spending so much time fixing bugs after production, they were now able to get to their enhancement list and create additional value.

> Rolling off too soon is a problem, but never rolling off is also a problem. Go back to the success criteria and ensure that all have been met or addressed before approaching the subject of rolling off the initiative and on to another effort.

Value = 10 ÷ (5 + 5) = 1 or 100% (Ideal)
Value = 8 ÷ (5 + 5) = 80% (Improved)

He again concluded that this team was positioned to take on the next challenge. They had dramatically increased the value they were delivering from roughly 30 percent to 80 percent. Getting additional investment dollars would give them the opportunity to increase their scope and the overall value of what they delivered to their clients. They were now positioned to win big!

Spencer finally evaluated communication and found it was exactly where it needed to be. Cross-functionally, the team was performing risk management, change control, and tweaking communication channels where appropriate. He assigned communication a value of 90%. This program, which was once in a position to *stop*, was now ready to go!

S.T.O.P. = Value + Team Performance + Communication
Go! 79% = (80% + 67% + 90%) ÷ 3

The only problem left for Spencer was convincing Jack he was ready to roll off...

A Better Way to Do Project Portfolio Management

The story told through the eyes of Spencer really focused on the use of our project management value equation (*Value*) in the context of a project or program. But could the same equation be used for project portfolios? How do executives, looking over a wide variety of products and projects in a company's portfolio, know when to stop or kill one? The decision to stop or kill a project amidst a sea of projects can seem almost impossible. Not only is there the question of *how* to stop it, but can enough information be gathered in time to know *when* to stop it?

The decision to *stop* a project or product line can't be made in isolation. It must be made in the context of the entire portfolio. Can we in the program management and project portfolio management community help our senior executives with this decision?

Let's stop for a minute to think about the concept and then further dive into the requirements/specifications of creating such a dashboard. What if there was a way for your senior management to see, in a one-page dashboard, updated in near real-time, how their portfolio of projects is performing? The dashboard would have to be simple and intuitive so that quick decisions could be made, but with enough data behind it to ensure the quick decisions were correct decisions.

The following concept is inspired by the recent advancements in technology and information gathering used by the financial and stock portfolio management community. The concept is called a "heat map." A heat map is a visual aid used to track large amounts of information very quickly to identify trends and make informed decisions. For stock trading, the concept is pretty simple; as stocks go up, the heat map turns green and shows the percentage increase for the day. As stocks trade below the previous day, the heat map turns red, showing the percent decline in overall value. Why can't this same concept be applied to program and project portfolio management?

To use a heat map for program and project portfolio management, you would have to slightly tweak the concept. For instance, you would need to weight each project differently based on the overall value it would bring to your organization. That value could be applied universally across all projects in the portfolio based on typical business case metrics like ROI or initial rate of return, or by total expected cost of each individual project. Each project's value would have to be displayed visually to provide the proper context regarding its relative value.

To do that, you would need to represent each project with an individual box or square (Figure 4.3). The size of each project box represents its relative value; the bigger the box, the bigger its value. Keep things simple by using only three box sizes to represent value. A simple high, medium, and low value determination will suffice. Value relative to size could then be kept in a legend at the bottom of the map.

Figure 4.3 Relative Project Size Map

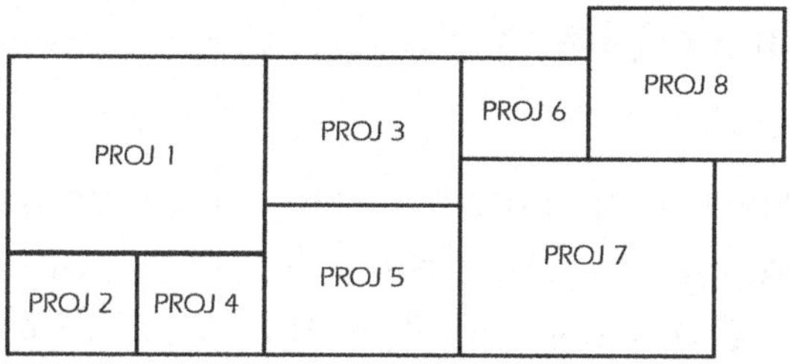

High ROI = \$20 million, Medium ROI = \$10 million, Low ROI = \$5 million

The benefit of this approach is that it visually represents in-flight projects with expected business value or ROI relative to each other.

Next, you would take your model and quantify the overall value of each one of the projects using the *project management value equation*:

$$Value = Scope \div (Schedule + Cost)$$

By assigning a value to each of these constraints, you are quantifying the value of the project. Increases to schedule and cost will decrease the value of the project if more scope is not being delivered to offset these changes. It's important to ensure that your scale for scope, schedule, and cost are the same. Assigning either a percentage value or using a 1–10 scale can work if you quantify each variable consistently. The idea is that at the start of the project, the overall value should be ideal and equal 100 percent. As tradeoffs are made between each of the three variables, the project value will decrease. Change to the project value could then be given a color to represent how the project is doing. For example, if any project has decreased in value by 15 percent, you could use yellow for that project on the heat map. This yellow representation would alert executives to projects headed for trouble and those where the fundamental underlying business case assumptions may be in jeopardy.

Projects decreasing in value by greater than 25 percent could be colored red, meaning that the project has gone off the rails and the original business case or ROI is no longer valid and must be adjusted. That adjustment could lead to the project being killed or postponed.

Having this ability to see project data relative to other projects would give executives better insight into how to balance their portfolios and manage resources more appropriately. Viewing data in this fashion also provides them focus; focus on where the trouble is and how to manage through it to maintain balance in the business and the portfolio.

Final Thoughts

If project management is to survive as a profession, then we as project managers need to realize that our job is more than just managing scope, schedule, and cost – it's about delivering value.

The iron triangle still exists; don't get me wrong. But to truly be effective as a project manager I believe a new triangle must emerge. A triangle based on managing value, team performance, and communication.

I developed the equations in the book to help identify and quantify the variables where more focus and management need to be applied on our projects. The equations work. However, I did not intend the equations to be prescriptive. My goal in writing this book was not about just providing new tools and methods for project managers. My goal was really to advance the profession through critical thinking.

I encourage you to use the equations in an effort to further develop them. See what works for you and where improvements need to be made. Let other project managers know what's working for you by generating further discussion about these topics. Stimulate the project management community and help each other to succeed.

In the story, Spencer was able to design a process, based on critical thinking that was applicable to his high-pressure project situation. He realized that although every project is unique, that uniqueness requires adaptability and a willingness to succeed. He used the S.T.O.P. principles as a way to create an adaptable, focused, efficient, and smart way to manage the team and ensure value was delivered.

You can do the same! The S.T.O.P. principles ensure that if your project is viable, it can be managed successfully. AND if it's not viable, you have the data, process, and communication tools necessary to *stop* it and *stop* it quickly!

Appendix

Figure A.1 The S.T.O.P. Card

- **Stop** before you act. Don't rush into a project without considering the risks and the benefits.
- **Think** about what you're about to do. What is the right way to safely accomplish the task?
- **Observe** the situation and surrounding environment. What are the risks? How can I reduce them?
- **Plan.** Develop your plan to reduce the risks and decide how to best implement the plan with safety and transparency.

Figure A.2 The Five Steps to Risk Management

THE FIVE STEPS TO RISK MANAGEMENT

1.**Identify hazards** to the project. Consider all aspects of current and future project situations, environment, and known historical problem areas.

2.**Assess hazards** to determine risk. Assess the impact of each hazard in terms of potential decrease in value based on probability and severity.

3.**Develop controls and make decisions** that eliminate the hazard and reduce the risk:

a.Reassess hazards given the controls
b.Determine the proper decision authority

4.**Implement controls** that will eliminate the hazard or reduce its risk.

5.**Supervise and evaluate.** Enforce standards and controls. Evaluate the effect of controls and adjust or update as necessary.

Risk management is a part of every project we do! It is the process of identifying and controlling hazards and making risk decisions to protect the value of the project.

Figure A.3 WBS PERT Analysis

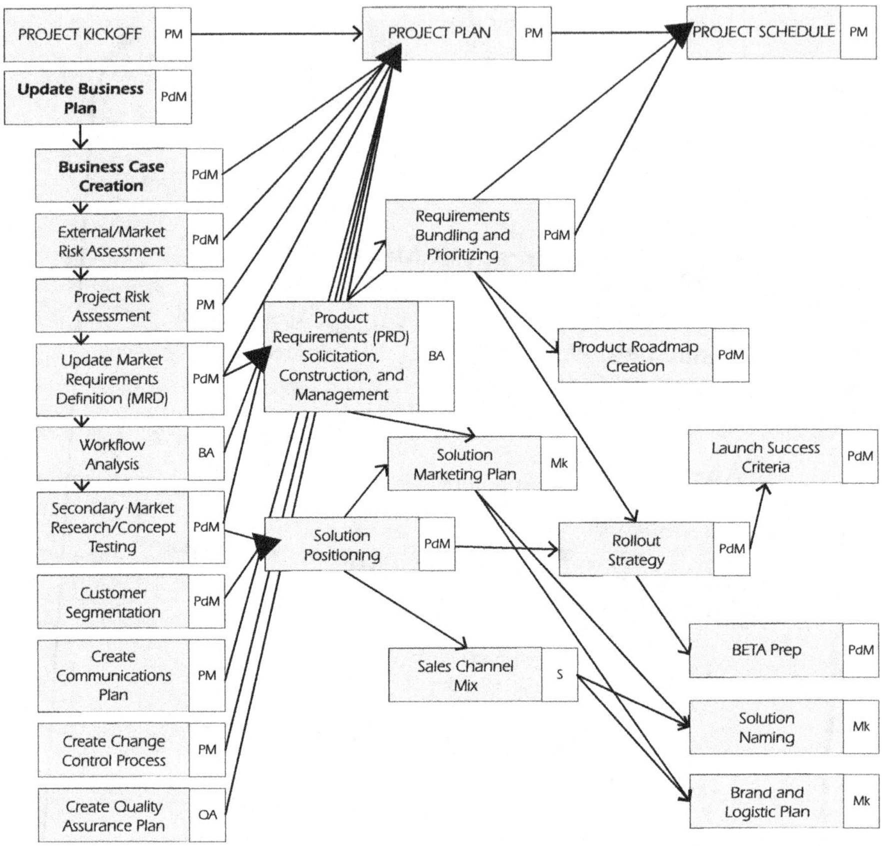

PdM = Product Management
PM = Project Management
Eng = Engineering
QA = Quality Assurance
Mk = Marketing
Ops = Operations
BA = Business Analyst
S = Sales

Index

R

S

T

W

X

Made in the USA
Monee, IL
13 September 2026

62981022R00125